The University of California Press acknowledges that this publication has been made possible, in part, by funding provided by the AMS 75 PAYS Fund of the American Musicological Society (AMS).

Automatic Artistry

CALIFORNIA STUDIES IN MUSIC, SOUND, AND MEDIA

James Buhler and Jean Ma, Series Editors

1. *Static in the System: Noise and the Soundscape of American Cinema Culture,* by Meredith C. Ward
2. *Hearing Luxe Pop: Glorification, Glamour, and the Middlebrow in American Popular Music,* by John Howland
3. *Thinking with an Accent: Toward a New Object, Method, and Practice,* edited by Pooja Rangan, Akshya Saxena, Ragini Tharoor Srinivasan, and Pavitra Sundar
4. *Key Constellations: Interpreting Tonality in Film,* by Táhirih Motazedian
5. *Just beyond Listening: Essays of Sonic Encounter,* by Michael C. Heller
6. *Making Stereo Fit: The History of a Disquieting Film Technology,* by Eric Dienstfrey
7. *The Composer's Black Box: Making Music in Cybernetic America,* by Theodore Gordon
8. *Automatic Artistry: Negotiating Musical Creativity in a Technological Age,* by Alyssa Michaud

Automatic Artistry

Negotiating Musical Creativity in a Technological Age

ALYSSA MICHAUD

University of California Press

University of California Press
Oakland, California

© 2026 by Alyssa Michaud

All rights reserved.

Cataloging-in-Publication data is on file at the Library of Congress.
ISBN 978-0-520-40576-9 (cloth)
ISBN 978-0-520-40577-6 (pbk.)
ISBN 978-0-520-40578-3 (ebook)

GPSR Authorized Representative: Easy Access System Europe,
Mustamäe tee 50, 10621 Tallinn, Estonia, gpsr.requests@easproject.com

34 33 32 31 30 29 28 27 26 25
10 9 8 7 6 5 4 3 2 1

To Chris

And in memory of Jonathan

Contents

Acknowledgments   ix

INTRODUCTION   1

1. THE PLAYER PIANO: FROM DOMESTIC MACHINE TO MUSICAL INSTRUMENT   19

2. THE REPRODUCING PIANO: CAPTURING AND RECREATING "LIVING PRESENCE"   51

3. SYNTHESIZERS: TECHNO-POP AND MECHANIZATION AS AESTHETIC   80

4. DRUM MACHINES: SYNTHESIS, SAMPLING, AND THE CULTURAL CONSTRUCTION OF AUTHENTIC SOUND   108

5. SINGING SYNTHESIS: AMATEUR COLLABORATION IN VOCALOID'S CREATIVE COMMUNITY   133

6. HOLOGRAPHIC PERFORMANCE: AUTOMATION AND PARTICIPATORY FANDOM   156

CONCLUSION   187

Notes   199

Bibliography   215

Index   227

Acknowledgments

Since my earliest childhood memories, writing has been the lens through which I have examined life, my way to connect with others, and a means of making sense of the world's complexities. Creating a book has been a lifelong dream, and turning that dream into reality has been a journey supported by the guidance, help, and encouragement of so many exceptional people. It is my privilege to be able to thank them here.

My first and longest-term thanks belong to my family. My mom and dad have encouraged me as a writer for as long as I can remember, and the legacy of their respective loves for music and technology are reflected in this book's very concept. I am especially indebted to my mom, who has supported me tirelessly during the most taxing stages of the journey, and am grateful also to Daniel and Carly for their love and care across these many years.

I owe my stepping-stone successes to each of my teachers, from the elementary school educators who invested care to cultivate my love for learning and the written word; to the undergraduate and graduate mentors who helped me to discern my path—notably Vernon Charter, James Janzen, Edwin Gnandt, Don Quantz, and Christopher Moore; to my doctoral supervisors, Jonathan Sterne and Lloyd Whitesell, who patiently worked to support me as I found my confidence and authorial voice. Jonathan passed away shortly before the completion of this book, but this entire project had its genesis when we rode in the back of a Montreal taxi, and he said to me, "How about automation technology?" The lingering fingerprints of his insights, style, and encouragement can be found in every chapter of this book.

I am grateful to the incredible team at University of California Press for supporting this project. Raina Polivka's generous insights and steady care have underpinned this long journey to completion, and I'm thankful for the ways in which she and editorial assistant Sam Warren have taken so much mystery and stress out of the process for a first-time author. My appreciation goes to production editor Emily Park and marketing manager Chloe Wong for their help bringing *Automatic Artistry* into the world. I would like to give sincere thanks to series editors Jim Buhler and Jean Ma for placing their confidence in this book. I also want to express my deep appreciation to the people involved with the FirstGen Program at UC Press, who illuminated each step of the publication roadmap and believed that my voice was worth amplifying.

This project benefited greatly from the constructive critiques of several insightful readers. My developmental editor Emily Doucet helped me retool tired old thinking into sharper new thinking and see the forest for the trees. Paul Sanden and David Suisman were my perceptive and generous peer reviewers, and their input has made this a much better book. Copyeditor Sharon Langworthy lent her keen eye to ensure that many small errors were caught and corrected. Rex Lawson helpfully shared his expertise on the player piano with me, helping to clarify important historical points. Seiya Okuda provided insightful interpretation support at NicoNico Chokaigi and clarified details about recent developments in Japanese Vocaloid culture.

Ambrose University has been a wonderful academic home for the past eight years, and its professional development funds have supported some of the travel and research assistantships for this book. I'd like to express enormous thanks to my colleagues in Ambrose Arts for their support. Mark Bartel held down the fort during the book's final stages and helped hold space for this project to succeed amid the ongoing demands of a small department. Barb de Bruyn and Baylie Adams cheered me on to the finish line. Val Lieske provided input on the early stages of chapter 6 and kept checking to make sure I was alive while I was writing. Linda Schwartz created the scholar-in-residence role that originally enabled me to complete and present research out of which the original seeds of this project grew.

I also count myself lucky to have terrific colleagues beyond Ambrose Arts. Thank you to Beth Stovell, Christy Thomas, Cindy Karikari, Colin Toffelmire, Darren Dyck, Jen Singh, Jon Coutts, Jonathan Lo, Kyle Jantzen, Matthew Morris, and Monetta Bailey for your support at many different times and in many different ways. The staff at the Ambrose library—particularly Patty Neufeldt, Megan Hallam, and Kim Kissack—have gone above and beyond to help me find obscure sources more times than I can count, and they continue to make Ambrose University an even better place to be a music researcher.

I would like to thank each of my students in the music department for their daily enthusiasm, insightful questions, investment in this book's progress, and hilarious sense of music-wing humor. It is a privilege to journey with each of you. Extra special thanks to John Glanville and Anna Konrad, who served as research assistants during the preparation of chapters 4 and 6, and also to the members of the winter 2025 edition of MU 403—Jackson Buller, Bella DeSouza-Cook, Anna Konrad, Joel Warren, and Bethany Wickens—who helpfully crash-tested early materials from the book's conclusion.

I feel incredibly grateful to have been supported by a community of hilarious, insightful, and caring friends who have sustained me in countless ways across this journey. Aiko Hosono, Brennan Eagleton, Dalton Keil, David Eddy, David Robles, Dawson Zimmermann, Deanna Smith, Elizabeth Clarke, Emily Wight, Jamie Boda, Jasmar de Torres, Jo Lo, Jordi Smith, Julio Pena, Kirsten Paige, Kristen Franseen, Kristen Prieto, Nelson Meira, Nina Penner, Rachel Finn, Ray Le, Rem Ritzinger, Seyren Xie, Stacy Duncalfe, Stephen Kan, and Teresa Davidson have discussed the ins and outs of this project across its many stages, sent along helpful discoveries, provided practical supports, served as thoughtful sounding boards, been essential diversions, and much more.

Barrett Hileman always defies categorization—colleague, department chair, mentor, and friend; artist, thinker, and writer. Barrett contributed innumerable hours of thoughtful input and encouragement at every stage of this project, and some of the best ideas in this book arose thanks to his storyteller's perspective, critical insights, and generous investment.

Finally, Chris Dyck, my muse since our student days, has been this project's biggest supporter, providing intellectual, personal, and pragmatic assistance every step of the way. After nearly a decade of reading, debating, listening, and cheerleading on this project, all I can say is: this one's for you.

Introduction

On a hot July evening in New York City, I joined two thousand people packing a Midtown auditorium for a pop concert. Decked out in brightly colored fan merchandise from past tours, the crowd sang and danced energetically to song after song. Midway through the show, the singer paused at center stage and called out to the audience, "Are you having a good time?" The crowd cheered in response. The singer egged them on, playing her part in a well-worn ritual between performer and audience: "I can't hear you!" Predictably, the crowd screamed even louder. But the irony of this exchange seemed lost on the audience. This performer's statement of "I can't hear you" was literal—she would never hear them, no matter how loudly they might cheer. That is because the singer was a hologram.

The singer's name was Hatsune Miku, and at this concert in 2018, she appeared onstage by means of projectors that cast her animated figure onto a large transparent screen that stretched across the front of the stage. Music journalists reacted to performances by Hatsune Miku with a mixture of derision, amusement, and incredulity, focusing on the features that her concerts lack when compared with performances given by a human singer. But this unusual performance was not an isolated technological stunt, and Miku was not the only hologram taking the stage in 2018. Concerts like the one I attended in New York were becoming increasingly common events, as an assortment of avatars made in the likeness of living, deceased, and fictional performers gave concerts in front of sold-out audiences. Virtual versions of Maria Callas and Roy Orbison also began tours in 2018, and the legendary Swedish pop group ABBA staged their painstakingly rendered digital comeback a few years later in 2022.

As digital performances have proliferated, some audiences have grappled with the questions they raise about the dynamics of the live concert experience. An audience member at the Salle Pleyel in Paris who had just viewed a Maria Callas hologram concert tried to articulate the conflict he felt as the audience struggled to negotiate its collective response to the digital opera star: "She comes on like a diva, waiting for everyone to stand up and scream, and there's some timid applause. People are wondering, 'Is this art? Is it serious? Do I get on board or not?' And we're captivated. It's scary."[1] The behavior of the Maria Callas hologram aligns with the way in which her creators anticipate the event will unfold, but she has no agency of her own. Her avatar walks out with preprogrammed confidence and pauses for applause she cannot hear or gauge. Her creators have made their best guess at appropriate timing and posture, but in the moment when Callas's hologram dips her head to acknowledge the audience, the crowd must negotiate their own collective response in real time; at the Salle Pleyel, the timidity of their applause reveals some uncertainty. If the singer is not present, are the audience applauding and cheering for the creators behind the animated avatar? Or are they suspending their disbelief and letting themselves be swept up in imagining a concert in which Maria Callas is physically onstage? The question, "Is this art?" reflects the fact that performances like those of the Callas hologram and Hatsune Miku seem to violate the established rules of what a live concert entails. A recital given in person by Maria Callas was easily understood as art, but what is a concert given by a preprogrammed Maria Callas hologram? What is a performance given by a completely fictional character like Hatsune Miku?

Musical performances that test the boundaries of what audiences understand as live concerts are not just a twenty-first-century phenomenon. For centuries, artists have used technological tools to interrogate definitions of creativity and explore the relationship between performances and audiences. While these questions are crucial as we seek to understand the meaning of technologically mediated artistry in our increasingly automated world, we will not find a complete answer by focusing exclusively on recent events, since the concepts and questions behind these boundary-pushing concerts have been asked as humans have probed the possibilities of automation technology in live performance for generations.

On another July evening in Germany, in 1926, nearly one hundred years earlier than the digital Hatsune Miku and Maria Callas concerts, an audience watched another performance that seemed to be missing a performer. A Welte-Mignon reproducing piano—an automated instrument with no human pianist seated at its keyboard—rattled off a dozen virtuosic pieces with mechanical precision, leaving the attendees unsure of how they ought to respond. One reviewer highlighted the inhuman capabilities of the machine, as well as the audience's hesitant reaction, which bore more than a passing resemblance to the response of the Parisian audience member in 2018: "The piano began to play . . . with an exactitude of which a human could never be capable, with a superhuman sonic force, with a geometrical clarity of rhythm, tempo, dynamics, and phrasing, which only a machine can produce. . . . The piano finished the composition and there was an uneasy pause. Should one applaud? There's no one sitting there. It's only a machine. Finally a quiet applause, growing louder. Calls of 'da capo.' And sure enough, the piano played it again, without hesitation, as precisely as the first time."[2] Although they are separated by almost a century, these automated concerts raised similar questions and prompted similar uncertainty from those who viewed them. During the "uneasy pause" in this reproducing piano concert, we can catch a glimpse of feelings and unspoken questions analogous to those of audiences in the twenty-first century.

Responses to these performances index intense debates about the nature of authentic musical performance and human creativity. Both the reproducing piano concert and the holographic avatar concerts played out before their audiences exactly as they were programmed to do by human artists, but the discourse around these technologies often centers on the worry that the introduction of automation technology into an artistic endeavor means that human connection and creativity will be erased, resulting in "soulless art" that may even fail to be art at all.

This book traces the cultural history of automation technology in music by examining 120 years of controversial performances and technological innovations, documenting recurring threads of resistance, optimism, and fresh approaches to creativity. Throughout this history, I show how automation is not a fixed horizon of fully automatic artistry toward which new technological advances move us increasingly close, but rather a shifting target, the objectives of which are heavily shaped by the cultural values of a given time and place. This flexible concept of automation serves as a revealing mirror that reflects cultural values, and for this reason, as an object of study it has the potential to bring artistic values into sharp focus, even across time and place.

Automatic Artistry tells the stories of three categories of automation technologies during the time when they were new, and in each story it relates the history of music lovers who embraced or resisted technological adoption in creative musical practice. We begin at the turn of the twentieth century, when player piano technology drew curiosity and concern as companies promoted "self-playing" instruments as one of a new array of early-modern commercial products that promised to simplify and enhance buyers' lives. The next stop on the journey visits pop musicians' initial experiments with synthesizers, sequencers, and drum machines in the 1970s and early 1980s. Here we delve into a heated cultural debate about the use of automation technology as an aesthetic choice that shed light on the implications of the emerging computer age. Finally, the book's third excursion moves into arenas of screaming Hatsune Miku fans armed with choreographed light sticks, where vocal synthesis software and avatar concerts in the 2010s provide a vivid opportunity to study the use of automation technology as a tool in the democratization of music making and the cultivation of musical communities. Across all three of these histories, I illuminate the process through which new music technologies move from messy uncertainty, anxiety, and resistance during their early years, to established functions and meanings that later appear to have been self-evident, after musicians and listeners have become comfortable with their use. This book shows how the ongoing process of folding new technology into musical performance facilitates expanded creative practices that ultimately amplify the artistic values of performers and listeners in fascinating new ways.

OF MECHANICAL DUCKS AND DRUM MACHINES

As strange as avatar performances have appeared to many music lovers, *Automatic Artistry* shows how the questions with which we are contending in the twenty-first century are the very same ones that have intrigued and frustrated observers for centuries. Of holographic singers and reproducing pianos a hundred years apart, listeners have stated with similar dismissiveness, "She is . . . only a hologram."[3] "It's only a machine."[4] These performances are part of a lengthy tradition of concerts that incorporate technologies that appear to eliminate human involvement, but upon closer examination, actually involve an intentional and artful concealment of the human effort that gives the impression of effortless automation. Similar performances (musical and otherwise) date back to the eighteenth century, including Wolfgang von Kempelen's chess-playing Mechanical Turk, in which a hidden human chess master operated a fraudulent automaton that seemed capable of defeating strong opponents in chess. A "self-moving machine" designed by Jacques de Vaucanson performed a dozen different tunes on the flute by means of mechanically blown air and a moveable throat and mouth; and a defecating duck, also constructed by Vaucanson, appeared to eat grain, digest it, and then relieve itself.[5] Decades after the defecating duck was first displayed, a viewer discovered that food swallowed by the duck did not in fact move into its stomach for digestion, but was held at the bottom of the mouth tube. This meant that fake excrement had to be loaded into the back of the duck before its digestive process was displayed. Jessica Riskin has observed that "this central fraud was surrounded by plenty of genuine imitation. Vaucanson was intent on making his Duck strictly simulative, except where it was not."[6]

Like the defecating duck and its predecessors from centuries past, avatar performances, drum machines, and reproducing pianos carefully emulate certain aspects of a musical performance—except when they do not. However, I argue that these departures from automated simulacra are not merely failures, or places where the technology is unable to live up to fully automated ideals, but the very places in which human creativity shines. Much of the challenge of parsing the relationship between creativity and automation is tied to the fact that creativity is often hidden from view at the moment when an automated process springs into action, seemingly without human involvement. In this book, I pull back the curtain on each technology that appears to substitute automation for creativity, from the holographic singers who never miss a note to the drum machines that play superhuman beats. However, by revealing "the man behind the curtain" in these stories, I am not seeking to expose a central fraud that punctures the illusion. Rather, the revelation of the human presence behind the technological curtain in these stories is a celebration of human ingenuity and an illumination of the new forms of artistic and technical mastery required of new instruments. My aim in *Automatic Artistry* is to illuminate the work of the artists who have charted new territory through the constantly evolving relationship between music technology, human creativity, and practiced skill.

Technological tools can give the impression of being solid and fixed, or of being self-evident solutions to issues that we retrospectively imagine, when the reality is that they are usually the highly flexible products of unfolding cultural dynamics. Technologies reflect their context in a particular time and place, and they also reflect the values of their makers and users, even—or especially—when the values of these two groups differ. This book's transhistorical approach highlights these shifting dynamics across time and place, showing how our fundamental curiosity about the relationship between technology and creativity is both a centuries-long human interest and a conversation that takes on new implications in each successive generation.

In the eighteenth century, writing about Vaucanson's mechanical flute player, Jean-Jacques Rousseau astutely observed, "It is not the automaton that plays the flute; it is the mechanic, who measured the wind and set the fingers in motion."[7] Behind the apparent ability of the wooden flautist, the player piano, and the drum machine to automatically create music is a human being who invests creativity and hard work. Vaucanson's cultural context inspired and contributed to his efforts to break down a flautist's mouth and tongue technique into mechanical motions. Though Rousseau was correct that it was not the automaton that was making music on the flute but its mechanic, I advance his argument one step further: Vaucanson, who invested intellectual and creative effort in the automated flute performance, was not just a mechanic, because just as much as a human flautist, he gave attention to musical factors such as tuning, rhythm, and pitch in the performance he crafted. Vaucanson made music, and was therefore a musician. So are the people whose stories are told in this book.

The first germ of this project and the research that followed dates back to 2011, when I saw news and videos of the first Hatsune Miku concert in North America. I was immediately curious, both about the meaning of an automated holographic performance and about the range of reactions the concert elicited. I spoke with people who rolled their eyes and spoke dismissively of Miku's fans as well as the performance, and with others who felt disturbed by the concept of a holographic performer and shuddered as they wondered aloud whether this was the beginning of the end for singers, or for live music altogether.

It was not until I took up these concerts in earnest as a research topic that I uncovered two insights that I found both wonderful and fascinating. The first was that the more I understood about Hatsune Miku and the Vocaloid software that synthesized her singing voice, the more clearly I could see that this new form of music making was not a threat at all, but a unique approach to creativity with its own aesthetic aims and cultural values that were rooted in a vibrant artistic community. The second discovery I made as I spent more time with the questions and fears surrounding the use of automation technology in musical performance was the staggering number of times we have trodden this same path with new music technologies, each of which have been painted as threats that spelled the demise of creativity or live music. From Sousa's depictions of the player piano and phonograph as a "mechanical menace" to the anti-drum-machine campaign by the Society for the Rehumanization of American Music; from the uneasy news stories about Hatsune Miku and Vocaloid to the present discussions around the use of generative AI; from the casual speculation about the future of creativity to the serious accusations of technological dehumanization—we have lived through this same story over and over again. In each subsequent generation, however, we seem convinced that it is our particular situation that is unique—that ours is the one that will strike the fatal blow to creativity as we know it. There is just one problem with this narrative: while critics are still busy debating their impending extinction, creative humans keep on being creative. They take up the very technologies that are framed as threats to their creativity and make new and exciting art.

One of the most harmful misunderstandings that circulates in debates about automation technology in music is the idea that new technologies threaten to directly replace vibrant human creativity with lackluster mechanical or digital replicas. These conversations—particularly in the popular press—often use terms that frame technology in the arts as a frightening conflict that is beyond our ability to control. "Creativity" and "Technology" become sentient entities that seem to have their own agency and will—Godzilla versus King Kong, battling to the death in a metropolitan skyline while we watch anxiously amid the rubble. This version of the story of technological innovation and human creativity is problematic not only because of the way in which it wrongly pits technology and creativity as opposing forces that cannot coexist, but because of the way in which it dehumanizes the creativity behind technological innovation.[8] Creative people create technologies that open up new artistic possibilities, while others use these technologies in fascinating and sometimes unexpected new ways. Technology cannot kill creativity (nor should we hope for creativity to conquer technology) when each of them is a tool that human artists with their own goals and agency can choose to apply to their work. Across *Automatic Artistry*'s technological and conceptual history, I argue that musicians' continued pursuit and use of automation as an artistic tool ultimately opens up new opportunities for creativity, rather than erasing it. This book's purpose is to reveal the new possibilities that have been infused into our processes of music making through the generative relationship between human musicians and their technological tools.

ALL MUSICAL INSTRUMENTS ARE TECHNOLOGICAL OBJECTS

In the twenty-first century, a course, a book, or a conversation about music technology is far more likely to focus on software like Pro Tools or hardware like DJ controllers than on innovations like electric guitars or Boehm's refingering of the flute. However, predigital musical instruments are just as much an accumulation of technological advances as their counterparts from the twenty-first century. The modern grand piano, with its cast iron frame and double-escapement action, reflects the significant technological developments of its time, and pianists have leveraged and absorbed these changes into their technique at the instrument. Even instruments that appear to be less a product of technological advances than the piano, such as the drum kit, have been improved to meet the needs of performers and ensembles. A trail of patents for technologies such as synthetic drum heads, mechanical kick drum pedals, and the more complex double-kick pedals reveals the drum kit's own history of technological development. As Emily Dolan has shown, studying musical instruments as technologies enables us to reintegrate an artificial distinction between the artistic and the mechanical and to investigate the values we have attached to musical tools that are often taken for granted.[9]

Many of the histories in this book track the reciprocal ways in which artists use new technologies to shape creative practice and, conversely, how musicians' creative practice shapes the final form of new technologies. Music technologies occupy complex social and cultural roles, which is the very reason they make a fascinating lens through which to examine changes in musical performance. What exactly is a music technology, then? Music technologies are tools with which musicians produce sound, and every technology in this book is indeed a tool that creative individuals take up out of a desire to create music. Music technologies are also sets of values that are embedded in material objects and digital processes, the design of which reflects a particular approach to making music. Music technologies are also artifacts that reflect specialized knowledge of music theory and technique, and they also in turn facilitate future acquisition of this knowledge and technique by other musicians. Finally, music technologies are products that are designed and marketed with an imagined customer and a set of imagined uses in mind, as well as a monetary price at which these functions are valued in a particular time and place. These complex and ever-shifting instruments of creativity are the results of an ongoing dialogue in which designers, musicians, and listeners codify the values of their cultural context.

At its heart, *Automatic Artistry* is a story of reciprocal influence between the values crystallized in technological processes and creative human users. There are times when the people whose stories are told in this book—as well as commentators in the popular press discussing today's latest technologies—use reified language to describe the outcomes of this dialogue, seeming to confer agency on music technologies when they say things like, "The drum machine changed the sound of pop music forever." These people likely do not intend to suggest that an idealized concept of the drum machine independently applied its own influence to "change" the way musicians create music. In many cases, comments such as this are a type of shorthand that encompasses the complex dialectical relationship between technological inventors, technological objects, and creative users. If a musician, producer, or listener says that drum machines changed pop music, this statement more accurately implies a broader landscape in which inventors' ideas, instruments' affordances, users' creative alterations, and contemporary aesthetic goals have swirled together in a complex and ongoing process of negotiation around synthesized and sampled percussion sounds. However, the instability that becomes apparent when any kind of music technology becomes the subject of a sentence instead of a human musician is especially evident when we are dealing with technologies of automation that are easy to describe as though they have agency of their own, despite the fact that they plainly do not.

The historical narrative of *Automatic Artistry* centers on stories of automation technology and musical performance, but in doing so, it also encounters technologies that seem to blur the boundaries between technologies that are automatic and those that are not. This is often because they simplify a complex process or enable musicians to execute techniques that would be beyond humans' capabilities without mechanical assistance. Kick-drum pedals and double-escapement piano action are technologies, and they are advances in instrumental design; however, they are not technologies of automation but of mechanization. This distinction is significant because these two terms are often confused in discussions about automation technology. A technology of mechanization is one that uses machinery to aid human effort, reducing the strength, precision, speed, or other physical requirement for a particular action. This does not mean that the mechanized motion is mindless or dehumanized. On a hammered dulcimer, an instrument that requires the striking of strings with hammers, a performer plays by holding a pair of hammers and using them to hit the strings. The intricacies of a grand piano's action enable far more rapid striking of strings with hammers, permitting a pianist to use ten fingers instead of two hammers to perform music far more complex than would ever be possible striking dulcimer strings by hand. Of course, the piano's mechanisms do not make it an automatic instrument. It will not make a sound without the pianist's physical action in the moment of music making, and pianists spend years mastering the intricacies and sensitivities of the mechanisms in their instruments.

Automation, then, is different from mechanization in that it carries out a predetermined process in the absence of real-time human interaction or intervention. A music box can be wound up and set down to play without ongoing human involvement. A sequencer can be programmed to perform a looping melody and will do so on its own indefinitely, until it is switched off or a component in the system fails. In many of the histories in this book, however, the people who are selling or using technologies conflate these two terms or apply them misleadingly. *Automatic* can be a vague adjective that is used haphazardly to imply some kind of reduction in human involvement. In *Automatic Artistry*, identifying the places in which music making is not actually automated is just as important as describing the situations when it is automated. In many situations when automation is applied as a descriptor, with a little bit of digging we find that these claims are either misleading marketing that oversells a device's simplicity or unfounded dismissals of musicians whose effort and creativity are intertwined with—but not erased by—a technological process.

A TECHNOLOGICAL TOPOGRAPHY

At the turn of the twentieth century, Western society was in the midst of a period of rapid technological and cultural change. The landmark Paris World's Fair in 1900 brought together culture and technology from around the world in a show of progress and optimism in the modern era. Thousands of electric lights adorned a twinkling Eiffel Tower, and visitors admired wireless telegraphs, moving sidewalks, automobiles, cutting-edge manufacturing technologies, phonographs, and many other technologies. Taken together, the displays at the World's Fair painted a picture of the future as a place of glittering efficiency, leisure, and interconnectedness, fueled by faith in human progress.

Alongside this faith and optimism in response to a changing world in which humanity seemed poised to remake their lives with these new devices, however, was a measure of technological anxiety and resistance. The very electric lights that shone on the Eiffel Tower at the 1900 World's Fair prompted a strike in New York seven years later, when lamplighters refused to light the twenty-five thousand gas lights that illuminated Manhattan, in protest against the arrival of new electric lights, the widespread installation of which would extinguish a career with a centuries-long history. Technological progress may have inspired hope and excitement at dazzling exhibitions, but for those whose occupations were impacted by the adoption of new tools, it also caused understandable apprehension. Fiction writers penned novels and short stories that examined similar concerns and questions, such as H. G. Wells's *The Sleeper Awakes*, Villiers de l'Isle-Adam's *The Future Eve*, and E. M. Forster's "The Machine Stops," which grappled with the relationship between technology and human identity and explored themes of technological reliance and dehumanization.

The phonographs that were played at the 1900 World's Fair were themselves just one innovation in a cluster of new technologies of sound recording, broadcast, and reproduction that contributed to a changing landscape for music listening in the early twentieth century. The phonograph, radio, player piano, and reproducing piano were discussed, used, improved, sold, and legislated alongside each other in an interwoven story of technological development in which the concepts of musicianship, liveness, and music listening were negotiated. For some, the opportunity to hear professional musical interpretations away from the public concert hall was a thrilling possibility that promised to democratize quality music and the development of good taste. For others, such as John Philip Sousa, the adoption of these devices threatened to cause the exact opposite: "a marked deterioration in American music and musical taste."[10] The definition of what counted as a musical performance came up for debate as companies worked to convince customers that their particular product was the one that was best able to transmit a musical performance by a human artist. However, the very terms of this debate were heavily contested. "Tone test" concerts featured live singers and Edison phonograph recordings side by side in an effort to legitimize phonographic reproductions as "real music," while reproducing piano companies insisted that their instruments conveyed the "living presence" of their artists.[11]

After examining these debates in the early twentieth century, *Automatic Artistry* visits the 1970s and early 1980s, when a dispute was once again raging around definitions of authentic musical performance. In the lead-up to the 1970s, the music industry was dominated by the sounds of pop, rock, and folk artists from the United States and Great Britain. Musical exports from these two countries came to epitomize popular music, even in other countries where English was not widely spoken. In the mid- to late 1970s, artists from a wider range of cultural backgrounds sought to work their way onto airwaves and record store shelves saturated by Anglo-American and British rock and pop, contributing to an increasingly diverse spectrum of sounds and styles. Alongside the continuing development of mainstream rock and pop, as the 1970s unfolded, developments in multitrack recording and programmable drum machines contributed to disco's tight, danceable rhythms and polished production, and synthesizers facilitated the development of distinctive sounds in new genres including the techno-pop of bands such as Kraftwerk and Yellow Magic Orchestra.

In 1971 Intel released the world's first microprocessor, and this innovation marked the advent of the computer age, sparking a flurry of new developments in computing technology. Researchers in the United States and Europe made significant advances in robotics, and these new technologies saw widespread implementation in manufacturing. Robotic processes were deployed in manufacturing with increasing frequency, and concerns about the implications of these advances seeped into popular culture discourse. Influential films including *Westworld* (1973), *Tron* (1982), and *Blade Runner* (1982) asked questions about the place of automation in modern society, and their soundtracks also reflected contemporary perspectives on electronic music. Fred Karlin's score for *Westworld* used electronic motifs to create a sense of artificiality, particularly in scenes in which the film's humanoid robots are framed as threatening or inhuman. Wendy Carlos's soundtrack for *Tron* liberally blended orchestral elements with synthesizers to reflect the movie's boundary-blurring commentary on virtual reality and creativity, while Vangelis's music in *Blade Runner* wove synth-rich soundscapes that evoked the film's vision of a dystopian future in which artificial intelligence had run amok. Soundtracks such as these made use of synthesizers as a sonic representation of the robotic age, reflecting a widespread conceptual coupling of electronic sounds and the technological future that was employed in genres beyond film music.

Worries about the substitution of automation technology for human labor had by no means disappeared in the second half of the twentieth century. In 1964 New York Congressman William F. Ryan proclaimed before the new National Commission on Technology, Automation, and Economic Progress, "Today we face such vast technological change in our economy that the question is asked, 'Is the age of automation a boon or a curse?' Is automation creating as many jobs as it wipes out? . . . Or is automation a 'Loch Ness monster,' swallowing up its millions of victims and then disgorging them into a sea of poverty-stricken idleness and despair?"[12] Musicians were not exempt from these worries about technologically fueled job loss, and fears about automation factored into discussions of drum machines, synthesizers, and recorded music in the 1960s and 1970s, when the American Federation of Musicians and the British Musicians' Union launched campaigns such as "Keep Music Live" and weighed the benefits of regulations for automated instruments.

The passage of another forty years in the story of technology and culture brings us to the 2010s, and to yet another moment in which optimism and anxiety, as well as the growth of new opportunities and a sense of loss around the old, intermingle in the changing musical landscape. The use of synthesizers and drum machines was no longer a bold artistic statement, and these twentieth-century instruments had become part of the twenty-first-century musician's expansive palette, which now included an array of new software packages, digital instruments, and affordable production tools. In the years immediately leading up to the 2010s, social media sites, notably the music-focused community MySpace, offered a new platform on which musicians could connect with fans and promote their work. MySpace included features for embedding videos on the newly launched YouTube video platform and a service through which artists could upload songs and albums, affording new artists greater control over their own work and promotional efforts.

In contrast with previous generations' pathways for building musical careers, many musicians in the early twenty-first century took advantage of affordable home recording equipment and the opportunities for self-promotion on social media to start their own careers and promote their own music. Australian songwriter Kevin Parker, best known for his self-recorded and self-produced project Tame Impala, launched a highly successful career working alone in his home studio, using an array of layered effects to create heavily textured psychedelic tracks, which he uploaded and shared through MySpace during the early years of his career. Even for bands that do not employ digital instruments and effects to the same extent as Tame Impala, this individualistic approach to production and use of the internet as a primary avenue for promoting music characterized many artists' origin stories in the early twenty-first century. Grammy award–winning folk musician Justin Vernon, who performs as Bon Iver, self-recorded his debut record not in a studio, but in a Wisconsin cabin. Vernon self-released his album and initially gained attention by sending copies to bloggers for online reviews. These approaches to recording and promotion took advantage of entirely new tools and media platforms to create music and reach audiences in ways that were impossible only a few years earlier, and they underpin the artistic stories not just of popular artists in mainstream genres, but of artists who used software like Vocaloid during the same time period.

It should come as little surprise that automation technology continued to be a prominent theme in popular culture in the 2010s. Perhaps more surprising, however, is that a TV series named *Westworld* (2016–22), and movies called *Tron: Legacy* (2010) and *Blade Runner 2049* (2017) were all released during this decade, directly linking the interests of filmmakers and producers of the 2010s with those of the 1970s, reviving the questions and stories that had occupied artists a couple of generations earlier. Other movies took up questions about artificial intelligence and the relationship between humans and machines, such as *Her* (2013), *Ex Machina* (2014), and *Automata* (2014), while the TV series *Black Mirror* (2011–) has tackled technological anxiety in many of its episodes, directly taking up questions about the ethics of holographic performers in a 2019 episode titled "Rachel, Jack and Ashley Too."

TRACKING TECHNOLOGICAL RECEPTION AND USE

This book spans more than a century of technological history, but our journey is not a jerky ride dictated by the disparate material details of player pianos and singing holograms. Rather, it is the human contexts of these technologies that map the route for travel, and our story follows the twin threads of the reception and use of automation technology in music. In order to access the human side of a technological story, this book draws on methods that facilitate a focused look at the stories of the people who bought, heard, worked with, objected to, adapted, and embraced these technologies.

The concept of *crisis historiography* in media studies describes the process of examining the shifting functions of a technology as it is shaped by users' perceptions and practices during the technology's early years.[13] As I apply it in this book, crisis historiography is an examination of a time of unstable "identity crisis" in which a new technology's meaning is both historically and socially contingent and is still undergoing a murky process of definition. Media scholars have advocated for the contextualization of historical media within their original networks of meaning, examining them not as "old media" from our perspective in today's world, but as "new media" within the context in which they were once new.[14] Both of these approaches ask us to consider the narrative that underpins the gradual solidification of new uses, features, and meanings of a technology, including those with which it was not initially associated.

In *Automatic Artistry*, chapters 1 and 2 are situated at the turn of the twentieth century and include in-depth engagement with print advertisements. These sources offer a glimpse of consumers' aspirations and attitudes, as well as the cultural realities they inhabited. In a study of early twentieth-century advertisements, Roland Marchand argues that although historians cannot know for certain whether any advertisements' messages were actually accepted by their readers, or even whether the content of advertisements reflects consumers' fantasies (or just those of the advertisers); ads are nevertheless historical sources from which researchers can make "plausible inferences" about popular attitudes.[15] This is due in large part to the fact that the writers who create ad copy are themselves expert communicators, and the success of their mission hinges on their ability to successfully understand and connect with their audiences. In the early twentieth century, consumers in the West navigated the transition to modernity in part by aligning themselves with emerging trends and fashioning their identities through their purchases. Advertisements from this time period carefully walked a tightrope between reflecting consumers' lived realities and painting a picture of the people consumers could become if they purchased the advertised goods.[16] Chapters 1 and 2 shed light on the way in which advertisements for player pianos and reproducing pianos reflected the musical sensibilities of potential customers and also appealed to their aspirations as creative and culturally refined individuals.

If the early twentieth century was an ideal time to focus on the attuning of the advertising industry to cultural values and economic aspirations, then by the mid- to late twentieth century, while advertisements were still useful sources in a people-centered approach to technology, new stylistic trends in music journalism also began to offer up a wealth of information through the detailed research and descriptive prose of writers in the 1970s and 1980s. Compared with the sensationalized reporting of the late nineteenth and early twentieth centuries, journalists in the 1970s and 1980s undertook significantly more in-depth reporting, transcribed the text of interviews in full with supplemental commentary, and wrote extensively researched sociological and psychological profiles of their subjects. Of course, journalists from this time period were by no means immune from bias, particularly when North American journalists interacted with music from overseas, as discussed in chapters 3 and 4. However, the quirks, assumptions, and stylistic tendencies of individual journalists in music magazines can be just as revealing as the direct quotes they capture from their interviewees.

The technologies and performances studied in chapters 5 and 6 benefit from a different approach. Situated in the early twenty-first century, the story of Vocaloid and its fan community is not one that can be fully revealed through advertisements or journalism. Vocaloid was not advertised widely during its early years, and the growth and development of the Vocaloid community generally flew under the radar of mainstream Western journalism for much of its history. When it did come to the attention of journalists in the mid-2010s, the critics who reviewed performances were often so unfamiliar with Vocaloid's cultural context that they wrote about concerts in dismissive or uninformed ways. The details of the ways in which creative users shaped Vocaloid's meaning during its early identity crisis years are best unearthed through fans' own words and actions, as well as the trail of digital discourse and art they left behind as they negotiated a contested technology through networks of creative art and online commentary. While journalism still factors into chapters 5 and 6, the methods for these chapters focus primarily on firsthand reports of experiences from people who are a part of the communities studied. In these chapters, I have also employed autoethnographic methods drawn from fan studies that have enabled me to take advantage of opportunities to view concerts, interact with members of the Vocaloid community, and participate in some of the rituals that have come to define Vocaloid culture.

A ROAD MAP FOR *AUTOMATIC ARTISTRY*

The first automation technology we encounter in chapter 1 of this book turns out not to be an automation technology at all. In the late nineteenth century, the player piano was originally marketed as an "automatic" device that could "instantly" and "effortlessly" produce perfect piano performances. In this chapter, I show how a great deal of advertising language and conceptual framing was borrowed from the labor-saving household devices (such as steam washers and sewing machines) alongside which these instruments were initially marketed and sold. However, the player piano—contrary to the messages in advertisements—was in no way instant, effortless, or automatic, and required significant practice and specialized technique to play well. The story of this technology is not merely one of an overhyped machine, however, but a dialectical development in which amateur musicians began to use this new technology in increasingly creative ways, companies responded by developing and promoting features on their player pianos that supported these objectives, and the player piano evolved materially and conceptually into a musical instrument in its own right.

At first blush, the electrically powered reproducing piano discussed in chapter 2 appears to be the fulfillment of the player piano's false claims of effortless automation, with companies claiming it could accurately reproduce performances by the great pianists of the day without any involvement from its owners. This chapter highlights the tension between our ideals of objectivity—and the way we seek to crystallize them in technologies of automation—and subjective human creativity. It tells the story of reproducing piano companies that invested in highly publicized efforts at developing scientific means of capturing the "living presence" of piano performances. Behind the scenes, however, the piano rolls that were sold with great pianists' names on them were less of an objective record of a performance event and far more of a creative collaboration between pianists, machines, and skilled human editors whose efforts were concealed in order to sell a particular idea of automation as a means of controlling and possessing human creativity.

Situated halfway between the decline of the player piano and the rise of holographic performance media, chapter 3 delves into the 1970s as a flashpoint in which the meaning and function of synthesizers in popular music was contested, and automation became an aesthetic in its own right, rather than merely a means of reproducing sound. By contrasting Germany's Kraftwerk and Japan's Yellow Magic Orchestra, I show how two of the most prominent electronic bands from this decade interpreted synthesizers in markedly different ways, each of which creatively blurred the boundary between human and machine labor and used automation technology to make an artistic contribution to the complex dialogue about Germany's and Japan's cultural identities in the wake of World War II.

In chapter 4, I take up the story of a contested automation technology that was frequently heard alongside synthesizers in the late 1970s and early 1980s. Out of all of the instruments in *Automatic Artistry*, the drum machine has the dubious honor of having provoked the most resistance and anxiety around the potential of an automated device to take the place of human musicians. This chapter traces the reception histories of instruments from 1980 to 1985, including the LM-1, which was the first drum machine to sample acoustic drums and was marketed with the contentious tagline "Real Drums." By drawing together the perspectives, stories, and creative outputs from the professional drumming community during the early 1980s, I show how the instability of the drum machine's artistic status during its identity crisis years resolved into an understanding of the drum machine as a new creative option with its own distinct merits and possibilities, rather than merely a replacement for a human drummer.

Chapter 5 delves into the twenty-first-century story of a vocal synthesis program developed by Yamaha called Vocaloid, which was intended to provide a way for studio professionals to mock up simple vocals for backing tracks. To the astonishment of its creators, who expected the software to sell just a few hundred copies, a dedicated community of amateur musicians and visual artists took up Vocaloid as an accessible tool for songwriting, and popular Vocaloid voice banks began to sell thousands, or even tens of thousands, of copies. In this chapter, I trace the explosion of the Vocaloid fan community online, showing how a group of creative amateurs succeeded in completely reinterpreting a new technological tool, creating a brand-new genre of music in which fans and amateurs collaboratively co-create Vocaloid music.

What happens when a fictional character whose music has been collectively produced by a global community moves from home computer screens to arenas full of screaming fans? In chapter 6, I visit Vocaloid concerts in North America and Japan, juxtaposing the dismissive confusion of journalists with an excavation of the rich layers of interpretive meaning and ritualized participatory interaction that fans have gradually developed at Vocaloid shows. This chapter takes the opportunity to carefully examine the questions that these concerts raise concerning the nature of live performance and its relationship to automation technology. I show that live performance and automation are not mutually exclusive opposites, but that at shows like Vocaloid concerts, the repetition and predictability that is facilitated by the use of automation actually paves the way for exciting new forms of creativity that originate in the audience rather than onstage.

. . .

Across the histories in this book, while the technologies change, the ingredients of the discussions change remarkably little. If the story traced in *Automatic Artistry* is not actually one in which increasingly sophisticated automation technologies are inching us ever closer to a world in which human artists are made redundant by machines, what can we conclude about the trajectory of our relationship with music technology? The best answers to these questions are not found in knee-jerk reactions to the latest technological identity crises. Studying the ways in which these conversations (and their attendant anxieties and misunderstandings) have played out in the past offers us critical insights into the way they will play out again in our future, as well as the ways in which we can respond most constructively to the changes that seem to destabilize our own musical practices and preferences.

1. The Player Piano

*From Domestic Machine
to Musical Instrument*

The first time I sat down at a player piano and tried my hand (or, more accurately, my feet) at pedaling through a piano roll, the results were, quite frankly, embarrassing. The pedals were cumbersome and did not respond easily to my efforts. Despite my best musical intentions as someone who has studied piano for many years, my tempo was unsteady, and the dynamic range I managed to produce was meager. I tried to build up enough air pressure to produce an accent on a climactic chord, and the result was little more than a thin wheeze. The reality was that I simply did not have the skill necessary for my physical technique to match my interpretive aims. In the years since that first encounter, I have had opportunities to attempt to improve my performance at several different player pianos, but the results, I am sad to say, have been dismal. The blame for these subpar performances rests not with my musicianship, nor with the machine, but squarely with my lack of skill at it.

My experience at the player piano not only frustrated my hopes of creating a musical performance of the piece I was playing, but it also sharply contradicted the prevailing cultural understanding of the player piano as an automatic device that can effortlessly turn out identical renditions of a piece of music. The most well-known representations of the player piano in popular culture since the middle of the twentieth century have painted the device as a benign mechanical jukebox that either plunks out familiar tunes for listeners to enjoy or else plays up the dehumanized elements of automated music making, framing the player piano's performance as an unsettling simulacrum of a human performance. Kurt Vonnegut's dystopian novel *Player Piano* (1952) took the name of the instrument as the title for a cautionary story of humanity's overreliance on technologies of automation. The television series *Westworld* (2016–22) featured the reproducing piano (a type of electrically powered player piano) in its opening credits and throughout the series to underscore themes of automation, control, and inhuman machinery. Executive producer Jonathan Nolan described the automated piano that sits in Westworld's saloon as "a 'primordial' version of the automated robotic hosts that populate the theme park in the series.[1]

Whether it was viewed as an automatic performer or a convenient musical playback device, the player piano was not the only music-making machine in the early twentieth century, and its punched-paper rolls presented just one vision for encoding musical sound. The player piano and the phonograph were developed and introduced nearly contemporaneously, but compared with the phonograph's vibrating diaphragm—a nineteenth-century technology that still sits in every home theater sound system speaker and every pair of headphones tucked in a backpack in the twenty-first century—the player piano's pneumatically powered machinery was ultimately relegated to the status of a historical also-ran. Once considered to be equal in cultural and technological importance to the phonograph, the player piano's role in the history of mechanically reproduced music faded almost to invisibility after the instrument's sharp decline in popularity in the 1930s.[2] This instrument is now a dusty mechanical curiosity with a split history, having been at times heralded as a democratizing force for the enjoyment of quality music and at others demonized as a threat to amateur music making.

In this chapter, I track discourses surrounding the player piano in its early years, revealing important changes in the representations of this technology in journalism, advertisements, music criticism, and instructional resources, which together chart a journey through the cycle of technological reception that this book illustrates. The player piano, initially introduced as an "automatic" and "effortless" device, borrowed its early conceptual context from household labor-saving devices such as sewing machines and steam washers, which advertisers promised would relieve human users from strenuous work. The player piano was similarly marketed as a device that could automatically perform music, and writers describing player piano concerts assigned musical agency to the machine itself. Two decades later, after a flurry of debates, changes to the devices' functions, marketing strategies, and publications, the player piano came to be understood not as a machine that could provide instant results, but as a musical instrument in its own right—one that provided mechanical support to music lovers in order to facilitate their individual creativity and enjoyment of quality music at home.

The reception history of the player piano is one in which issues of creativity and automation were renegotiated by copywriters, critics, and amateur musicians, ultimately turning the meaning of a new music technology on its head. The resulting reconfiguration of the relationship between humans and machines across the player piano's history calls into question the meaning of automation technology in music: Where is the line between mechanical aids to creative human music making and dehumanizing automation? Each of the technological stories in this book tests this question in a different way, and the case of the player piano approaches it from a valuable angle, not only because of the way in which the capabilities of the device were so dramatically overstated in its early years, but also because of the way in which musical amateurs are situated at the center of the story. Many technological histories advance a narrative that links the stories of high-profile professional work with the unfolding history of a new technology, but amateur users are frequently given less sustained attention. Player piano scholarship includes a great deal of work on professional composers' relationships with the instrument—including those of Conlon Nancarrow and Igor Stravinsky—but studies of amateur users have been infrequent.[3] However, amateurs were the most common users of both the player piano and the piano itself. During the nineteenth century, the piano had become an essential purchase for any family who considered themselves to be a part of the growing middle class.[4]

Despite the ubiquity of the piano at the turn of the twentieth century, hundreds of thousands of player pianos were sold between 1900 and 1930. In fact, in the years following World War I, sales of the player piano surged, and the number of units sold each year eventually surpassed that of traditional pianos.[5] How were these instruments marketed and sold when many customers already owned pianos? While there was no immediate consensus on what the player piano was supposed to do, and for whom, advertisements, magazine columns, books, and other sources offer us a window into this period of potential and transition by showing how multiple interpretations of the player piano jockeyed for acceptance in the minds of critics, potential buyers, and owners, before the instrument's material configuration and meaning were solidified.

PIANO-PLAYERS, PLAYER PIANOS, AND REPRODUCING PIANOS: A LEXICON

The term *player piano* as many people use it in the twenty-first century often serves as a catchall to refer to three different categories of machines. In the late nineteenth and early twentieth centuries, each of these types of devices was discussed separately, and each category was understood as a product that performed a distinct function. The earliest of the three

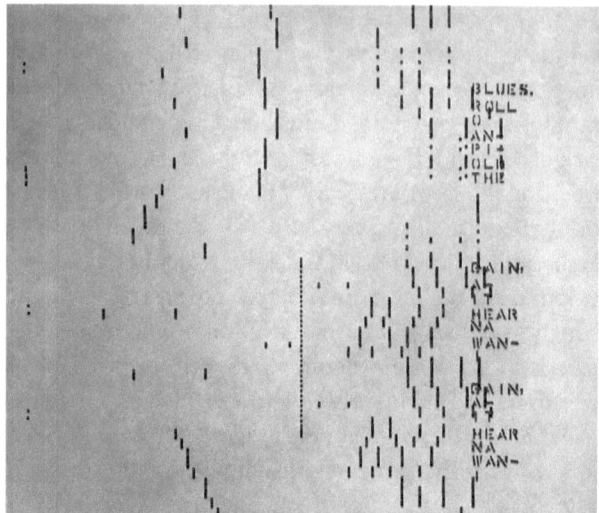

FIGURE 1. Punched-paper piano roll. Each hole represents a note struck on the piano. As the paper unrolled during playback, the song's lyrics would become visible, facilitating a sing-along.

devices was the *piano-player*. It was joined in the twentieth century by the player piano, and finally by the reproducing piano, which entered the market shortly afterward. Each of these machines operates using piano rolls—perforated sheets of paper that encode the information needed for playback in a series of punched holes (see figure 1)—and all three types are used to perform piano music by means of pneumatic systems. Beyond this basic similarity, however, each machine is distinct in its form and function.

From the late 1890s through the early 1900s, the Angelus Piano Player and machines like it were sold separately from pianos (see figure 2). Since they were not built into a piano cabinet and could be rolled up to any keyboard, they could indeed play "your" piano, as the ads claimed, or any other piano in front of which they were placed. At the time, these machines were called *piano-players*, and this term accurately described the function of the device. With its mechanisms encased in a wooden cabinet approximately the same height as a piano keyboard (see figure 3), the piano-player could sit directly in front of any piano, occupying the space which a human pianist's body would ordinarily occupy. In this position, a set of horizontal mechanical fingers extended over the piano keyboard—in the same space that a human performer's hands would rest—with each of its mechanical

FIGURE 2. Exterior cabinet of the piano-player. *Source:* William D. Parker and Edward H. White, *Automatic Piano-Player*, US Patent 592,641, filed April 5, 1897, and issued October 26, 1897, 1.

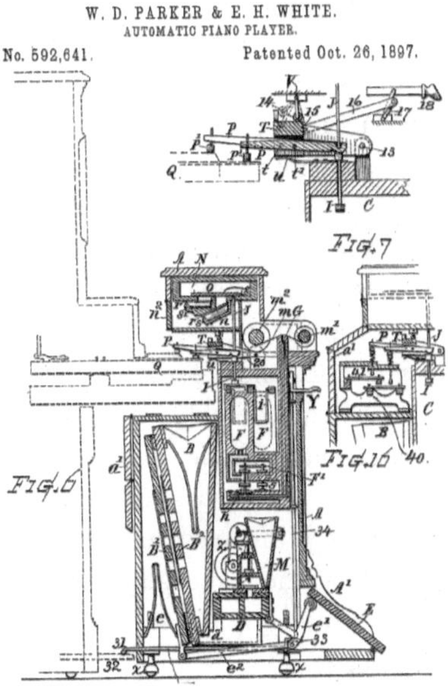

FIGURE 3. Side cutaway diagram of the piano player. *Source:* William D. Parker and Edward H. White, *Automatic Piano-Player*, U.S. Patent 592,641, filed April 5, 1897, and issued October 26, 1897, 3.

FIGURE 4. Woman operating the Angelus piano-player. *Source:* Nordheimer, advertisement, *Toronto Evening Star*, January 22, 1900, 2.

fingers positioned over a single key. Once the piano-player was set up in this way, the user could then sit down behind the device and pump a pair of foot pedals in order to power its pneumatic systems (see figure 4).

The term *piano-player*, then, did not identify the human user, but rather the purpose of the device, which acted on, or "played" a piano. Calling this device the piano-player puts it in the company of other technologies named after the people who originally performed the same task, such as dishwashers, governors, and computers. In many ways, the mechanical piano-player replaced the human piano player's body when its cabinet sat in the performer's place in front of the keyboard and extended its own fingers over the keys.

Piano-players were initially marketed as automatic and self-playing devices, but in reality this description was far from the truth. Although performers did not need to supply pianistic technique at a piano's keyboard when using a piano-player, they were still able to make decisions regarding important musical elements of the performance. The performer's hands operated levers that controlled the tempo, the balance between treble and bass, and the damper and soft pedals on the piano. The feet, meanwhile, provided power, but by varying the speed and depth of each pedal stroke, they also directly impacted the dynamics, phrasing, and accentuation of a performance. Unskilled pedaling would result in a performance that sounded considerably worse than merely neutral or expressionless playback. If an automatic device is one that executes a preprogrammed sequence of events without user involvement, then the constant effort required, the user agency involved, and the wide range of potential outcomes that were possible at the piano-player show that this instrument was not automatic at all; it merely mechanized certain elements of the process of music making. Coaxing a satisfactory performance out of a piano-player with all of its levers and pedals still required practice and musicality. An excerpt from a comic poem penned in 1919 captures the agony of listening to a lackluster performance on a player device from a neighboring apartment:

> Yea, bitterly cuss the sarcophagus ghoul
>   Who chauffeurs with murderous fin
> Insane permutations of sad syncopations
>   Accented, I'd say, on the "sin."[6]

Given the varying permutations rendered in this overheard performance, it seems that the performer in question lacked good pedaling technique! If the piano-player were truly an automatic device, the user's skill level would have been irrelevant, and the instrument would have turned out identically correct performances time after time without intervention, rather than the "insane permutations" the poem's author overheard.

Player pianos, the second category of instruments to emerge onto the market, were sold alongside piano-players starting in the late 1890s and began gaining popularity in the early 1900s. Unlike the push-up piano-player cabinets, the all-in-one player piano was a complete instrument with built-in player mechanisms included inside the body of a fully functional piano. The term *piano-player*, then, no longer accurately described its function or its material configuration, and these instruments became known as player pianos. The only significant difference in the machinery of a piano-player and a player piano was its location. Both instruments required the exact same technique and skills. They were still powered by foot pedals, and performers still controlled their own dynamics and tempo with hand levers and pedaling technique, for better or for worse. By 1905 player pianos had surpassed piano-players in popularity, owing to the convenience of being able to choose between playing at the keyboard unobstructed or making use of the player mechanism without rolling a cumbersome machine up to a separate piano keyboard.[7]

The difference in name between piano-players and player pianos is significant because it describes not just a design change, but also the way in which the instrument is understood. In the term *piano-player*, the word *player* is the noun, and the entire term specifies the actions carried out by the player machine. In other words, the piano-player is a device that plays the piano, much as a pencil sharpener is a device that sharpens a pencil. In *player piano*, on the other hand, the word *player* is an adjective that merely describes a type of piano. This shift from piano-player to player piano subtly strips away the agency that the earlier term (and much of the marketing for it) conferred on the instrument and leaves the control and authority in the performance instead to the human user who is playing the player-piano. These changes were not merely superficial rearrangements of words and mechanisms, and each one lends itself to a competing understanding of the instrument. In its early years around the turn of the twentieth century, the meaning of this technology was still uncertain. Was it a labor-saving tool meant to replace a human? Or a new type of piano that reconfigured the work of making music? The difference between a large machine that could be positioned in front of a piano in the place a pianist would have sat, and a machine that was hidden away inside the piano cabinet and left the pianist's bench unoccupied, is critical. The different configuration altered the relationship between the piano and the musician who sat down in front of it. Both devices provided the exact same mechanical assistance. The player piano, however, eliminated a machine barrier between human and instrument, and its name lent itself to the idea that users were merely playing a different type of piano, rather than using a machine to play the piano for them.

In 1905, M. Welte und Söhne was the first company to begin marketing the reproducing piano. The reproducing piano was different from the piano-player and the player piano in its use and purpose, and the messages of its promotional material were distinct as well. These instruments were built to automatically reproduce the exact musical details of a particular performance, and to do so without variation. They were electrically powered, not pedal powered, and did not require any physical or interpretive effort from their owner during the automated playback process. Piano rolls for reproducing pianos included data for dynamics and tempo changes punched directly into the paper, and these elements of the performance were also automatically incorporated during playback. Advertisements claimed that these instruments reproduced "the playing of the masters *at their best*. No subtle tone is lost; no shade of feeling unexpressed."[8] During the player piano's heyday, reproducing pianos were a costly luxury item, which sold only a fraction as many units as their pedal-powered counterparts.[9] And yet the reproducing piano, now often simply but inaccurately called a player piano, is the type of player instrument most commonly depicted in literature, film, and television, owing to the uncanny allure of the performerless performance as the keys move up and down on their own. I revisit the reproducing piano, as well as the implications of a fully automated playback system, as the focus of chapter 2. In the story of the player piano, however, the reproducing piano serves as an important conceptual contrast: it was an automatic device on which one could flip a switch and walk away while a performance unfolded with no further intervention or effort from the user. A comparison with the reproducing piano reveals just how misleading it is to describe the player piano as an automatic instrument. This misunderstanding, however, is not merely a twenty-first-century failure to understand an early twentieth-century technology. The piano-player and player piano were described as "automatic" in their own time, by parties with a particular interpretive agenda in mind, and this feud over labels is an important part of the historical negotiation of this instrument's purpose, as well as the larger debate about the meaning of automation technology in musical performance.

THE PIANO-PLAYER AS A LABOR-SAVING DEVICE

The piano-player, perhaps unsurprisingly, was initially understood primarily through the lens of its relationship with the piano itself. In the century leading up to the invention of the piano-player, keyboard instruments including pianos and small organs were increasingly found at the center of domestic life.[10] Owning a piano was a marker of social status, and daughters in middle- and upper-class families frequently received lessons.[11] The cultural context of the piano during this century was shaped in part by a growing list of music journals and circulars that included well-known titles such as Britain's *The Musical Times* (published since 1844) and the American publication *The Musical Courier* (1880–1961), as well as magazines that focused specifically on the piano, such as *The Etude* (1883–1957). These publications contain a wealth of information regarding the role of music, the piano, and piano lessons in nineteenth-century Western society, but they also offer a broader context than the strictly musical by situating musical instruments alongside some of their overlooked counterparts.

At its inception in 1880, before *The Musical Courier* adopted this more familiar title, it was named *The Musical and Sewing Machine Gazette*, followed by a stint as *The Musical and Sewing Machine Courier*. Sewing machines were eventually dropped from the title, leaving the publication with the name under which it was known until 1961. As this publication's early title suggests, however, pianos and sewing machines were frequently bought and sold side by side, owing to the fact that they were both marketed to women for use in the home. Salesrooms in the late nineteenth century displayed both sewing machines and keyboard instruments, and traveling salesmen hauled both types of products in their wagons between rural communities.[12] A small number of fascinating devices even took this coupling one step further and combined both a reed organ and a sewing machine in a single cabinet.[13]

The piano was also sold alongside a wider range of domestic appliances other than sewing machines. Retail stores in the late nineteenth century also offered household technologies such as toasters, freezers, and wringing machines alongside keyboard instruments.[14] The *Journal of Domestic Appliances and Sewing Machine Gazette*, which was printed between 1877 and 1930, detailed for its readers the latest improvements to devices such as coffee pots and knitting machines; gave brief descriptions of new patents granted; and advertised the latest products, from steam washers to typewriters. Many of these advertisements share a common theme: they focus on the labor-saving features of their products. An ad for Greenall's Steam Washer, which its makers called "The Great Domestic Labour-Saving Machine," introduced it as the "most efficient, easiest, and quickest Washer made. Accomplishes in Two Hours what is now a Hard Day's Work. Washing Day made a Pleasure."[15] The language used in this steam washer ad is typical of many of the ads found in the *Journal*. Across advertisements for everything from washing machines to button-hole machines, descriptors such as "automatic," "efficient," and "simple" promote products that promise to make domestic tasks quicker and easier to perform.

The *Journal of Domestic Appliances and Sewing Machine Gazette* included not only devices used for household chores but also keyboard instruments, which were also advertised frequently, and discussed their developments alongside those of the labor-saving household devices. In the January 1886 issue, numerous descriptions of patents that provide "improvements to the pianoforte" appear in the "Applications for Patents" section, and an article offering advice on renting out pianos and sewing machines describes a course of action for situations when a rented piano or sewing machine was accidentally destroyed or damaged while on the renter's property.[16] Although a musical instrument might appear to have had little in common with devices for completing household chores, the piano itself—much like washing and wringing machines—was a tool with which women were expected to carry out domestic responsibilities. Young women in the nineteenth century were given piano lessons partly to improve their appeal to potential suitors, but once they had married, women continued to play for their spouses, and men's expectation of being able to listen to their wives at the keyboard is reflected in literature, art, and instructional materials in the latter half of the nineteenth century.[17] If the piano was viewed as a household item with which these women performed services for their families, then these instruments made sense alongside washing machines and sewing machines in terms of their function and market, as well as the fact that pianos themselves were heavily mechanized pieces of machinery.

Mechanization—the incorporation of mechanical assistance into a process that would otherwise be executed manually—is not an "all or nothing" attribute with a clear dividing line between human and machine performance, but a process that proceeds little by little across the history of a technology. Harry Braverman has singled out the point in a technology's history at which a mechanism regulates the path of motion taken by a tool—even if that motion is not automated—as an important juncture in the technological evolution of a machine.[18] The sewing machine, with its pedal-powered needle following a fixed path up and down, is taken as an example of this machine-regulated motion in Braverman's work. In the late nineteenth century, however, sitting alongside sewing machines on sales floors and on the back of traveling salesmen's carts was a machine Braverman didn't discuss. The piano, with its complex systems of hammer and damper action, had taken advantage of regulated motion and increased efficiency in the mechanized production of sound long before the sewing machine was invented.

The piano is one of the most complex mechanized instruments in the music world. It is a musical instrument, but it is also very much a machine. As a particularly intricate machine (even peculiarly intricate, when compared with the simpler instruments that have historically been most prominent throughout much of the rest of the world), the piano itself is, as critic Ernest Newman notes, a "record of an incessant piling up of mechanism."[19]

Newman argued that violin players have achieved their high levels of expressivity by improving on human body parts with additional tools: the invention and use of the bow served in effect to elongate the fingers and soften their pressure. The piano, too, is essentially a mechanical dulcimer, with complex pieces of machinery designed to strike pieces of wire more efficiently than the human hand could do alone.[20] Just like washing machines and writing machines, the piano's mechanisms are more consistent and require less effort to achieve greater accuracy and speed, owing to the fact that the tool—in this case, the hammer that is striking the string—is no longer held and manipulated by a human hand, but by a mechanism. The piano's history as an increasingly complicated machine has significant implications: it means that player piano technology did not constitute a shocking shift from purely human musicianship to soulless mechanical imitations. These instruments were merely another step in the "incessant piling up of mechanism" that was the piano's inheritance. Like the piano itself, the player piano was both an instrument and a machine, and the two classifications were neither opposites nor mutually exclusive.

Given *The Journal of Domestic Appliances and Sewing Machine Gazette*'s focus on mechanized, labor-saving devices, and its inclusion of pianos, piano-players were a natural addition to its pages. Early piano-player technology appears in the *Gazette* long before it was advertised in newspapers. The *Gazette*'s writers—like most authors describing brand-new technology in the earliest stages of its cycle of reception—described it in terms that convey a sense of newness and awe. An article from 1887 recounts an early "International Sewing Machine and Domestic Appliances Exhibition," at which labor-saving products from manufacturers around the world were displayed.[21] A significant number of musical instruments were included in the exhibition, including early versions of Aeolian organs (which also played back notes from perforated paper rolls), and even singing dolls that could produce various tunes. The author refers to a device called the "automatic piano," introducing it with little fanfare alongside the Aeolian organ: "The automatic piano is a splendid instrument, which can be converted into an ordinary piano by merely removing the automatic arrangement. Another excellent invention is the Aeolian organ which can either be played by a skilled player by hand, or by one who has never before seen an organ, if he merely presses down the pedals. A perforated tune sheet is inserted at the back of the keyboard, to render the instrument automatic."[22] From this description, we cannot ascertain exactly what the "automatic arrangement" was, or even who was exhibiting this instrument, though from the mention of the fact that the automatic piano can be converted into an ordinary piano, it seems this mechanism was similar to the devices that would appear on the market a decade later. Of additional interest in this write-up is the description of the organ, in which the author focuses on the claim that a completely untrained user could operate it effectively.[23] This emphasis would parallel themes similar to early piano-player ads, which touted the devices' ease of use and quick results and conjure in a reader's imagination an automatic device that completely removed the necessity for skill and practice. Taken together, descriptions of nineteenth-century labor-saving devices provided a basic vocabulary from which early piano-player advertisements developed their own closely related refrains less than twenty years later.

IDENTITY CRISIS: THE PIANO-PLAYER IN EARLY ADVERTISEMENTS AND ARTICLES

"This Will Play Your Piano," proclaimed *The New York Times*'s first piano-player advertisement on May 1, 1898 (see figure 5).[24] Beneath an illustration of a boxy device that resembles a black cabinet, the advertisement describes the Angelus Orchestral Piano Player as "a new and wonderful invention that converts any piano into a self-playing instrument. No musical talent required to operate it. Instantly applied to any piano—grand or upright."

There is no piano depicted in this advertisement, nor is there a user shown operating the piano-player. The rectangular unit neatly conceals the mechanical action, which the ad promises will do the work of playing the piano instantly, with no skill, musicianship, or practice necessary. The bold heading "This Will Play Your Piano" suggests that it can do so in the place of a human musician. The trouble with these promises, however, was that they were largely false.

The year 1898 was pivotal in the history of the piano-player in the United States. During this year, two companies—Wilcox & White and Aeolian—played a key role in the introduction of piano-player technology to the readership of American newspapers from coast to coast by publishing their first regular advertisements in major papers. from the *Chicago Daily Tribune* and *The New York Times* to the *Minneapolis Tribune* and the *Los Angeles Times*. This year is of particular interest in the piano-player's reception history because it offers a close look at the way in which these companies chose to introduce their products during their earliest months of marketing. By 1899, the number of piano-player ads in American newspapers had exponentially increased, as companies undertook increasingly aggressive marketing campaigns, and new companies added fresh competition to the market. In the span of just a year, newspaper readers went from seeing

FIGURE 5. Early advertisement for the push-up Angelus Orchestral Piano Player, 1898. *Source:* Wilcox & White Co., advertisement, *New York Times*, May 1, 1898, 20.

advertisements for piano-players for what was likely the first time, to seeing the technology promoted on a weekly or even daily basis.

The themes of automation and ease of use found in the ad for the Angelus in figure 5 are common to most of the ads from 1898. Similar to other ads from that year, a Pianola ad from October 2 states, "When placed before any kind of a piano, [the Pianola] will play it with humanlike effects, rivalling the performance of a great pianist." This advertisement, like most early promotional copy, attributed the act of creating a musical performance to the machine itself, rather than to its human operator. The difference between the claim "*This* will play your piano" of the earliest ads and the idea "*You* will play your piano" is crucial. These advertisements assign musical agency to the machine, not to its potential buyer or user. Unfortunately for those who purchased piano-players believing the claims that the machine would instantly create quality musical performances, however, piano-players required significant practice. As the comic poem excerpted earlier in this chapter suggests—and my own experience confirms—while mindless pedal-pumping could coax sound out of the instrument, this nominal type of engagement would produce decidedly unsatisfactory results that did not "rival the performance of a great pianist" in any meaningful way beyond the insensitive pressing of a similar sequence of keys.

While most of the 1898 piano-player ads position this new technology as an automatic performer, they occasionally go one step further, suggesting that the automatic performance might even be superior to a human performance. The first Pianola advertisement to run in *The New York Times* makes even more grandiose claims than the Angelus advertisements that had recently run in the same newspaper: "It gives the most marvelous reproductions of solo playing, but it does not stop there. It will execute things which no performer could dream of executing. Many of these effects the human performer would like to produce if he could. In some respects, therefore, its feats are in the direction of ideal music. It has further powers and possibilities hardly yet suspected."[25] Here again, if the reader took the advertising copy at face value, they might believe that the machine itself was independently creating these effects and executing these feats. But what exactly were these vague "things" that the human performers wished they could execute, and what hidden "powers" did this product possess? The writers of this ad notably chose to describe the capabilities of the Pianola in only the most elusive of terms. By leaving the details to the imagination of a reader who likely had no experience with this brand-new technology, the device could be given a larger-than-life quality. This tactic was especially significant given the technology's state of "identity crisis" early in its reception history. The piano-player itself did not yet have established patterns of use. Was it meant to quickly provide acceptably tasteful music when entertaining guests? Or to push the envelope of pianistic possibilities? Was it simply amusing, or educational, or was it poised to replace physiologically limited human pianists? In a manner that has much in common with twenty-first-century advertisements that similarly overstate the potential of brand-new technologies to change our lives or perform things they cannot actually do, these piano-player ads from 1898 leaned toward the realm of fantasy and hyperbole.

While advertisers often play a key role in shaping the discourse regarding new technologies in their early years, they are not the only voices influencing the discussion. At the turn of the twentieth century, articles that discussed the possibilities of piano-player technology also appeared in music magazines and trade publications. The earliest of these approached their subject matter with an explanatory tone that was shot through with a sense of excitement and wonder that is typical of the early stages of the technological reception cycle. In 1899, Robert Braine opened his assessment of piano-player technology in *The Etude* magazine with this exuberant appeal to a historical figure for an imaginary endorsement: "If Father Bach could . . . see an innocent-looking wooden contrivance . . . begin to play one of his most difficult fugues or sarabandes with perfect accuracy, without a human hand being touched to the keys, I am convinced that he would raise his arms in amazement to the skies and declare that the age of miracles had revisited the earth."[26] This article mirrors some of the technological enthusiasm present in early accounts of exhibitions. But while Braine's article, simply titled "Self-Playing Pianos," is eagerly optimistic in places, it also provides a contrast to early advertisements by offering a detailed and pragmatic discussion of this new instrument.

In his description of the piano-player, Braine counters some of the fantastical language of the early advertisements by underscoring the fact that a person without musical training and skill could not play these instruments to their fullest potential: "While it is, of course, true that a child can learn to manipulate the machines in a few minutes, yet at the same time it can readily be seen that no one but a highly educated, practical musician can manipulate it to the best advantage, since the pedaling and tempo depend on the performer, and in the case of the pneumatic machine, the pianos and fortes as well."[27] Braine's account offers one of the most realistic assessments of the piano-player in the late-nineteenth century, laying out clearly for the musically trained readers of *Etude* magazine the fact that the additional mechanization of the piano-player could not compensate or substitute for a lack of musical sensibility. This early appraisal underscores the reality that the piano-player could not in fact make musicians of nonmusicians, nor could it enable someone with "no musical talent" to produce a coherent musical performance.

Braine's article addresses two other points that are notable for a piece published in 1899, and in doing so it anticipates the same anxieties and hopes that music lovers will wrestle with in subsequent stories in *Automatic Artistry*: fears around potential loss and hopes around expanded access. Braine first tackles the specter of technologically fueled job loss, noting that "some music teachers" think that the piano-player will "injure their business," believing that music-loving parents will be inclined to buy a piano-player instead of paying for piano lessons for their children. Writing in 1899, Braine can only speculate about the future, but his outlook is optimistic: "It seems to me that this idea is far fetched, however, as I should think that children of a family who possess one of these automatic pianos would be very much more anxious to learn to play than those of a family who do not. It might as well be claimed that frequent concert-going would take away the desire to learn music, whereas we all know that the opposite is true. It is more likely that this invention would increase the business of the teacher by increasing musical knowledge and intelligence among the masses." In addition to addressing the concerns of piano teachers, Braine's article comments on the opportunities afforded by this developing technology, focusing especially on the ways in which music lovers could access more of the musical repertory with the help of the piano-player. Noting that the educational opportunities would be especially meaningful to a family who lived a long distance from musical centers, Braine writes, "What a wonderful privilege and pleasure it is for a music lover who is not a performer to be able to step into his drawing-room, place his machine at his piano, with its faithful wooden fingers above the notes, and the next minute to be reveling in the sublime measures of Beethoven's "Egmont Overture!"[28]

Unlike the 1898 advertisements, this article does not overstate the perfection or artistry of these machines' renditions, and Braine notes clearly that the piano-player cannot substitute for the touch of a trained human pianist, but he is not attempting to argue that the two are comparable or interchangeable. He explains, "Its greatest use is, of course, for home entertainment and pleasure, and also for the dissemination of knowledge of leading musical works by making it possible for the non-musician to be able to produce them in a fairly intelligible manner, at least so that when he hears them properly performed at concerts, he will be able to appreciate and comprehend them." In other words, the piano-player offered rudimentary access to a realm of music that might have otherwise been gated behind time-consuming training, concert hall locations, or endless ticket purchases. Robert Braine's account of the piano-player offered a balanced consideration of a fledgling technology's future that did not resort to hyperbole or speculative fantasy. Even this reasonable approach, however, could not fend off the next stage of the piano-player's reception cycle: the technological panic.

MECHANICAL MENACES: THE PIANO-PLAYER AS A THREAT

The piano-player's history did not unfold in isolation, but was part of a much longer lineage of automata and mechanized innovations. While music boxes and other automated mechanical novelties existed as far back as the thirteenth century, the eighteenth century was a golden age for automata, an era in which viewers delighted in the machines and artists drew on their capabilities to explore questions about the nature of life. Alongside the automata devised by Jacques de Vaucanson that fascinated audiences in the mid-eighteenth century, Joseph Haydn also wrote thirty-two pieces for the mechanical clock organ between 1772 and 1793, and Mozart's K.608

stands as an example of a late eighteenth-century work for an automated instrument (also a mechanical clock organ) that leveraged its medium to explore themes of death, time, and the mechanical sublime.[29]

While mechanical musical instruments were still popular throughout much of the nineteenth century, Carolyn Abbate has summarized a shift in cultural attitudes toward automated mechanical devices that unfolded across the lead-up to the twentieth century, revealing a slowly growing cultural fear that these machines had the potential to rob humanity of the uniqueness of their abilities.[30] Emily Dolan and John Tresch have written of the symbiotic assemblage of human and machine elements in French opera in the 1830s, at a time when machines were objects of both fascination and scorn, vessels of the transcendent and job-threatening monsters.[31] Describing the ambivalent attitudes toward musical automation technology at the threshold of the twentieth century, Abbate writes, "By 1900 marionettes and automata, vast music boxes, and music machines with their phantom hands, are all Janus-faced, both magical and terrible."[32] In the case of the piano-player, too, depictions from the 1870s through the 1890s focused largely on the promise and potential of the technology, but in the years immediately following the turn of the century, as technological anxieties surfaced and debates grew sharper, optimistic interpretations were joined by criticisms that forecast a terrible side to these devices.

Between 1901 and 1906, the initial curiosity of early commentators and the overstated ads for the piano-player gave way to an increasingly intense conflict, one with much higher stakes. An article printed in *Musical Opinion & Music Trade Review* in January 1903 used strongly militaristic imagery to frame the growing popularity of the piano-player as a harbinger of mechanical music's impending conquest of the arts:

> This [piano-player] vogue is now regarded—and rightly regarded—as one of the most significant phases in the life and advancement of this mechanical age. It is heralded by the enthusiastic as a portent of the dawning of a new epoch, when machinery will still be the motive power of civilisation, but will be applied to uses hitherto deemed sacred from its invading banners. In other words, these persons . . . regard as now at hand the day when mechanical inventiveness will invade the precincts of art and will fix its ensign in the very altar of that domain. If this vision is to be realised, the piano player will certainly be the most prominent factor in its accomplishment.[33]

By evoking images of invading banners and a conquering army fixing its standard to a sacred altar, this author stoked the fire of the very same fear-fueled, binary debates that continue to rage in the pages of twenty-first-century periodicals: the fight between technology and human creativity in the arts.

While this *Musical Opinion* article does not anywhere come out in condemnation of the mechanized invasion it predicts, other writers were more than willing to emphatically denounce the new devices. John Philip Sousa famously villainized the phonograph and piano-player in his 1906 article, "The Menace of Mechanical Music," in which he warned in no uncertain terms that new technologies were poised to destroy human creativity and musical ability. "Sweeping across the country . . . comes now the mechanical device to sing for us a song or play for us a piano, in substitute for human skill, intelligence, and soul," Sousa wrote.[34] Arguing that the incursion of mechanically reproduced music would result in "a marked deterioration in American music and musical taste, an interruption in the musical development of the country, and a host of other injuries to music," Sousa maintained that the mechanically recreated sounds of the phonograph and piano-player were diametrically opposed to the authentic musical expressions of the human soul. In a similar vein to many of the early piano-player advertisements, Sousa's descriptions often attribute agency to the machines themselves, painting technological devices as having desires and beliefs of their own: "The host of mechanical reproducing machines, in their mad desire to supply music for all occasions, are offering to supplant the illustrator in the class room, the dance orchestra, the home and public singers and players, and so on. Evidently they believe no field too large for their incursions, no claim too extravagant."[35] An illustration provided for the article by F. Strothmann echoes this attribution of malicious machine agency, depicting a pair of phonographs and a piano-player lunging forward, wild-eyed, toothed mouths gaping, perforated paper billowing behind the lurching piano-player (see figure 6). "Does it go about to seek whom it may devour?" asks the caption, which links the technology with demonic adversaries through an evocative reference to a biblical warning: "Be sober, be vigilant; because your adversary the devil, as a roaring lion, walketh about, seeking whom he may devour."[36]

Alongside dramatic comparisons to military invaders and devils, Sousa sounded an alarm about the dangers of mechanical devices substituting for human skill and artistry, and in this anxiety he was not alone, nor was he speculating about a purely imagined trend. For several years, many public concerts featuring new player instruments had been touted by their respective companies using language that diminished the role of human performers. In 1904, promotional descriptions of a concert that included collaborative performances with the Triumph Piano Player introduced the

"Does it go about to seek whom it may devour?"

FIGURE 6. Illustration of a menacingly anthropomorphized piano-player and phonographs, featured in Sousa's article "The Menace of Mechanical Music," *Appleton's Magazine* 8 (1906): 284.

concert in language that seemed to erase human agency: "A grand concert is to be given by the Triumph player in St. James's Hall on March 11th. The Triumph will accompany Mr. Ben Davis, Miss Elizabeth Parkins, Miss Margaret Thomas, and Földesy. The Triumph and Ashton's selected orchestra will perform the complete Mendelssohn Concerto in G minor (Op. 25). The principle [sic] features in this player—which make it possible to undertake a task like this—are flexible aluminium fingers (which give a touch exactly like the human hand) and the direct stroke from the pedals (which enables instant accentuation of notes or chords)."[37] According to this description, the concert would not be given by a human performer, but "by the Triumph player." The writing suggests that the machine itself would be performing

Mendelssohn, and that it, and it alone, would be accompanying the other musicians. Of course, the reality that is not stated here is that an unnamed, unmentioned human musician would have been operating the Triumph. We can know this for certain, since the Triumph is specifically described as having pedals that enabled the user to instantly accentuate certain notes or chords, revealing that this was not an automated instrument. But the individual performing on the Triumph at this concert received no credit for this effort, even though it would have been a considerable effort indeed, requiring both musicianship and practice. Of course, this write-up was meant to promote the Triumph, and it is therefore understandable that the copywriter focused on the machine, but the erasure of the player-pianist in this concert, and many other concerts like it that were advertised in the late nineteenth and early twentieth centuries, seemed to imply to the reader that a human musician had indeed been replaced by a machine. Advertisements such as this one, which preceded the concert itself, give no indication of the quality of the results the piano-players delivered in concert, but this matter was taken up by other commentators during this same period. Many of them were dismissive of the musicality of the results obtained from the use of piano-players, but not all.

In a November 1901 opinion piece in *Musical Opinion & Music Trade Review*, a writer using the pseudonym "Anti-Pianola" complains: "I naturally feel a little jealous, being a pianoforte player myself, in listening to what appears to be a fine performance of a classical piece by a person with musical feeling, but who cannot even play a hymn tune; whereas I have spent so much time and trouble in acquiring a technique much inferior in quality and dexterity to that displayed by the above named instrument."[38] Here, Anti-Pianola suggests that the piano-player provided an unfair shortcut to users who did not actually possess the piano technique that the device appeared to confer on them, and further implies that the results obtained with the piano-player were better than those that a student of the instrument could attain through study and practice. These comments prompted a response from another reader, T. G. Dyson, whose letter was printed in the following month's issue. This contributor, however, did not write in to affirm Anti-Pianola's assessment or to validate the complaints in the earlier letter. Rather, Dyson first notes having tried "all the different kinds" of player instruments on the market, then summarizes their strengths and weaknesses, and, most significantly, casts suspicion on the veracity of the testimonials provided by accomplished pianists who claim that people with no musical talent can work them successfully. Dyson notes—much the same as Robert Braine did in 1899—that the machine will not make a successful musician out of someone with no musical sense: "Sir,—Like your correspondent 'Anti-Pianola,' I was interested in the development of the piano player, but in a little different respect.... As your correspondent states, pianists naturally feel jealous, and I have been surprised to read the high flown testimonials purporting to come from them. The piano 'player' can do many things that a good pianist cannot, and the pianist can render some pieces as no 'player' will ever do. However there is yet another straw left, for no one can hope to work successfully any piano player unless he has some music in his soul."[39] These comments acknowledge that it is a reasonable and instinctive response to feel jealousy toward a new technology whose makers claim it can exceed the skill that one has trained long and hard to acquire. Dyson astutely observes, however, that given the instrument's mix of strengths and weaknesses, the testimonials that are "purporting" to come from pianists are highly suspect. Furthermore, by questioning the testimonials of great pianists, Dyson is also deftly questioning the veracity of Anti-Pianola's self-disparaging assessment. Anti-Pianola's comparison of the piano-player's outputs and a trained human pianist's technique reveals a remarkably impoverished perception of the relative quality of musical performances. Furthermore, the choice to use a pseudonym was uncommon in this publication. If the name "Anti-Pianola" was taken on in order to conceal affiliations with Pianola or bias in the company's favor, much like the testimonials of professional pianists, then the anonymous writer's strategically performed inferiority seems to be little more than a ploy. Though it may not be possible to discover Anti-Pianola's true identity well over a century after these comments were printed, Dyson's measured and musically informed comments reveal Anti-Pianola's words to be an ignorant opinion at best and an attempt to plant tactically manufactured envy at worst.

Across these early years of technological anxiety surrounding the piano-player, the most insightful voices in the debate were not those who stoked fears about a technological takeover of human artistry, nor those who breathlessly praised new innovations as miracle-working wonders, but musicians themselves, who had the expertise to discern the strengths and weaknesses of various applications of this technology as it developed. Dyson's assessments were echoed by musically trained authors in other periodicals, who provided measured assessments and selective praise that focused on the unique opportunities this technology could offer, rather than stoking fear about what it might replace. In an article titled "Mechanical Music" that was published in *The Musical Times* in 1905, William H. Cummings, who at the time served as the principal of the Guildhall School of Music in London, reminded fearful readers that "there is no new thing under the sun."[40] Cummings, who was a baroque specialist, excerpted a 1737 letter from Sarah, Duchess of Marlborough, in which an automated musical instrument excited intense delight from the royal family, with the duchess describing the machine as "the finest piece of music that was ever heard" and noting that "Handel and all the great musicians say it is beyond anything they can do." Cummings notes that the memoirs of Handel and other musicians of the time did not reveal despondency or worry that this new contraption was going to sweep away the value of their own abilities, and that any anticipated decimation of harpsichordists and organists did not occur. "There need be no fear that machinery will ever displace the combination of brain, soul, and finger," Cummings writes, in a closing statement that stands in defiant contradiction to Sousa's contrasting prediction of the displacement of a similar trio of "skill, intelligence, and soul."[41]

At the core of many technological anxieties across this book is a worry that the latest new development in music is a frightening first step in the dehumanization of art, erasing one of the very things considered most central to our human identity. Even when time has eased these fears as they relate to individual technologies, the fact that they recur, and the intensity with which they do so, is a powerful indicator of where our values lie. Criticisms of the piano-player and player piano often point to an underlying belief in the value of hard work in the development of musical craft. The technologies examined later in this book reveal the cultural values of their own times and places, and taken together, studying the criticisms, perceived threats, and conceptual instability around new technologies in music gradually uncovers a fuller picture of how our most closely held values are inscribed in musical performance.

CREATIVITY AND AGENCY: THE LATER YEARS

Across our historical journey in *Automatic Artistry*, technological panics are regular occurrences. In the midst of these panics, the loudest voices are typically not the voices of experts like Braine, Dyson, and Cummings, who consider new musical tools with balance and thoughtful consideration. However, what the calmer voices of disciplinary experts often offer, in addition to the reassurance that human creativity is not on the brink of extinction, is the suggestion of what a new technology might become, beyond a mediocre replacement for traditional performance.

Once the technological panic subsides, and once music lovers have had the opportunity to assess the affordances of a new technological tool, a mechanized or automated instrument can come into its own, distinct from the comparators to which it was tethered earlier in its life cycle. Taken together, the advertisements and articles examined thus far depict the piano-player when its meaning was still uncertain, and when the device was often inaccurately discussed (and feared) as an automatic replacement for a pianist. Within a few years of the turn-of-the-century rush of anxiety around this technology, two significant shifts took place. First of all, self-contained player pianos began to overtake the push-up piano-players in popularity. At the same time, journalists, salespeople, musicians, and player piano companies began to use new language to describe player devices, and the conversations around these instruments in the 1910s and 1920s encouraged their use in increasingly artistic ways.

This emerging understanding of the player piano as an expressive instrument in its own right was advanced by a number of writers in books and periodicals, especially in the 1910s and 1920s. One of its most well-known proponents was British musicologist and critic Ernest Newman. Perhaps best remembered for his four-volume work *The Life of Richard Wagner*, Newman also authored a 187-page book titled *The Piano-Player and Its Music* in 1920. This book was published as part of a series called The Musician's Handbooks, which included other books such as *The Complete Organist* and *Memorising Music*—titles that were clearly aimed at performing musicians and not just people who were interested in automated playback on a machine. Given the educational nature of the other titles in the series, a reader might expect that *The Piano-Player and Its Music* would offer instructional content, but Newman's contribution to the series does little in the way of teaching hopeful piano-player users how to make music with their device.

Newman's book opens with a chapter titled, "A Defence of the Piano-Player." Certainly *Memorising Music* does not begin with a chapter penned in defence of memorizing music, nor would most other books discussing an instrument need to begin by asserting the validity of their subject matter. Newman admits early on that "for a musician to put in a plea for the piano-player in these days is to make a good many worthy people doubt his sanity or his honesty or both."[42] Newman wrote this comment, not in the late nineteenth century, when the device was still new and unknown, not alongside other hyperbolic articles during the technological panic shortly after the turn of the century, but in 1920, near the peak of the player piano's popularity, when player pianos were outselling nonplayer pianos. Newman remarks early in the book that people had hinted he must have been paid by "the makers" to support the player piano, but in response to the naysayers who objected to the player-piano on the grounds that it produced "mechanical music," Newman asks, "Where, in truth, *is* the non-mechanical musical instrument?" Arguing that humanity has been gradually improving the quality of the mechanical components of their musical instruments since moving beyond the unaccompanied use of the voice to perform music, Newman notes that the increase in mechanical quality has produced music of steadily increasing quality as well.

Responding to critics of the player piano who contended that "the performer has less to do with the making of the music than the hand-pianist has," Newman builds his argument on the idea that music making and technique are entirely separate matters, suggesting that the time spent practicing mindless exercises to achieve strength and independence in the fingers does not train musicianship and artistic sensibility.[43] With a player piano, the need for manual skill at the keyboard can be bypassed, and the performer can make use of mechanical assistance to lighten the load of technique necessary to apply their musicality to more complex musical works. From this perspective, the player piano begins to sound less like a dehumanizing technology and more like a mechanical upgrade, with more in common with flute keys and pipe organ bellows than automata and music boxes. The piano-player and the player piano, although they were initially framed in advertisements and other writings as "automatic," required a significant measure of real-time user control during the performance, as opposed to a truly automated machine that would run in the absence of any ongoing intervention. In fact, as player piano technology improved and companies introduced new instruments, this artistic control was gradually expanded, not diminished. Throughout the 1910s and 1920s, player pianos were frequently equipped with additional buttons and levers that gave users the ability to shape the finer artistic details of the piece in line with their own creative ideas.

This development is particularly evident in player piano advertisements run after 1910. A Kranich & Bach advertisement in *American Homes and Gardens* in 1912 anticipates some of Newman's arguments about the separation of technique and musicianship, claiming: "Only the technique–the striking of the right notes at the right instant–is automatic. Every phase of musical-expression is under absolute *personal control of the performer.* And 'expression' is what makes music–not technique. . . . Among the many exclusive features of superiority, one of the most important is the TRI-MELODEME or TRIPLE SOLO device, which enables you *personally* to 'bring out' the melody whether in the bass, tenor or treble, and subdue all else."[44] An ad by Sohmer & Co. in 1913 takes this type of rhetoric a step further, painting a fanciful picture of musical freedom, competence, and artistry for its readers. Beneath a heading that insists, "You PLAY the Sohmer Player Piano—You Don't Merely 'Operate' It," this same ad explains, "The expression devices of the Sohmer Player Piano are the opposite of mechanical. When you play the Sohmer Player Piano the expression and feeling in the music which you produce are your own—produced directly by yourself." In these later ads, gone are the references to labor-saving devices and ease of use. New verbs like *play*, not *operate*, underscore the idea that this machine is also a musical instrument, and those who play it are musicians.

In the 1910s, salespeople were also encouraged to carefully frame player pianos as instruments. The guidelines and tips offered to player piano salesmen in trade publications from the 1910s show how dealers were urged to avoid using words that brought machines to mind, focusing instead on artistic possibilities.[45] A 1911 issue of the trade magazine *Player Piano* suggested: "As far as possible, the mind of the purchaser should be diverted from the idea that he is buying a piece of mechanism. . . . The player piano should never be referred to as a 'machine'. . . . By the same token the player instrument should never be referred to as a 'self-player' or 'automatic'. . . . Never use the word 'operator' when referring to one who uses or demonstrates a player piano. A person operates a sewing machine or a lathe . . . it requires intelligence and musical taste to play a player piano, and such a person is a 'performer' as much as one who uses the fingers."[46] It is no coincidence that this advice specifically mentions that the player piano should never be compared to sewing machines and lathes, when in fact it was these very machines with which early piano-players, and even pianos themselves, were frequently grouped on exhibition floors and in the press. Advertisers and salesmen in the 1910s were selling player pianos not as automated music-making devices but as instruments that could facilitate satisfying and high-quality musical performances for people who lacked keyboard skills—provided that they committed to practicing them.[47]

MASTERING THE PLAYER PIANO

If Ernest Newman's book provides a rigorous defense of the player piano as a musical instrument, then Sydney Grew's comprehensive instructional guide, *The Art of the Player-Piano: A Text-Book for Student and Teacher*, offers its readers a detailed method for attaining proficiency on it. Published in 1922, *The Art of the Player-Piano* is not a brief pocket guide or slim method book; Grew fills more than three hundred pages with a thorough description of the finer points of player piano technique. The first sentence of the book's preface provides a brief summary of its entire central point: "The Art of the Player-piano lies in the pedalling and in the use of Tempo-control Lever or Buttons."[48] This opening statement echoes a timeless piece of musical advice: there is no substitute for the fundamentals. Musicianship at the player piano is elevated from a mindless exercise to an art form not through the use of the latest knobs and switches, but through skillful pedaling and sensitive tempo control. Throughout this book, pedaling remains at the heart of player piano technique: "Pedalling is as breathing in singing or fingering in pianoforte playing," says Grew in the book's opening pages. "The player-piano, like the pianoforte and the organ, is a musical instrument; its control is an art, and the performer an artist. . . . The instrument is to be stimulated, not driven. It is to be made to operate, not by crude physical force, but by movements induced by musical feeling and guided by musical knowledge."[49]

Sydney Grew's book is divided into two sections, with the first focusing on comprehension of the basics of musical form and structure, the use of levers, and studies in counting. The second section, which occupies twice as many pages as the first, delves deeply into the skills necessary for effective pedaling and their relationship to prosodic meters. As a textbook, this method integrates not only instructional writing, but pieces "for close intellectual study," to which the author expects the reader to attend with focus and persistence: "The intellectual effort required . . . is slightly less than that required in the study of instrumentation, canon, building construction, algebra, and so on; but it requires the same qualities and similar determination."[50] The chasm between Grew's perspective and the early advertisements that claimed anybody could seat themselves at a piano-player and be instantly and effortlessly able to produce a quality musical performance can hardly be overstated. Grew recommends a systematic approach to musicianship in addition to technical competence. He suggests that students first pedal through a piece no less than twelve to fifteen times in order to familiarize themselves with it, followed by practice reconstructing its form, rhythm, and patterns of accentuation from memory. Clearly, no instant or easy results are expected here. While Grew claims it takes approximately seven years for someone to be considered a good pianist, organist, or singer; for the player piano, he states, "It takes about three years to make a good player-pianist of a man or woman of average musical intelligence."[51] According to Grew, the player-pianist would attain a level of basic skill with the pedals after approximately a month of effort, becoming proficient at pedaling easily and quietly, controlling dynamics, and altering the tempo with sensitivity.

What came next, if a player pianist had attained these skills? Even more nuanced pedaling. Above all, as books and player piano magazine columns tell their readers, this is where the performer had the greatest control. A skilled pedaler could place a single pedal stroke at just the right moment, to draw on just the right amount of stored power, in order to shape the phrase exactly as desired. Good pedaling needed to be economical, be gentle, and match the cadence of the piece. Grew describes the pedals of the player piano with language that is at turns almost reverent and sensual, comparing them to the tools of skilled artists and to delicate systems in a living body:

> The player-pianist caresses the pedals. He controls them as a driver controls high-spirited horses. He transmits to them the subtle spirit of the movement which music sets dancing through his soul. He employs them with the delicacy the sculptor employs his tools; he also hews with them, as the woodsman drives into the tree with his axe. He treats them as the conductor treats his baton, marking not only the beats, but the rhythm also, and effecting phrasing, tone contrast and quality, climax, and the thousand and one details of effect—objective and subjective—which go to the making of musical performance. The Pedals are as the centre of a nerve system, from which are radiated commands to every part of the instrument—the most delicately intimate, as well as broadest and most sweeping.[52]

The attention Grew devotes to the use of the pedals in his book is noteworthy because it underscores the fact that learning to play the player piano well was a long-term investment, not simply a matter of purchasing a device that had certain novelty functions. Pedaling was a skill that required long-range planning and was not a matter of applying more pressure at the exact moment of a note or chord that the user intended to be louder. Grew notes, "The foundation of the art of player-pianism is establishing and maintaining elastic firmness in the pedals."[53] The effort needed to maintain this firmness would change throughout a piano roll, with sparse textures requiring very little power, and dense, chordal passages consuming far more. Skilled users would be aware of these shifts and adjust the strength of their pedaling accordingly, in addition to continuing to pedal appropriately for the rhythm, meter, and accentuation as needed.

Evidence from periodicals in the 1920s suggests that some consumers who purchased player pianos did end up following advice like Grew's, and practiced on their instruments enough to obtain satisfactory results. From May 1925 through January 1931, a column titled "Player-Piano Notes" was published in *The Musical Times*, providing readers with reviews of recently released player piano rolls. For the most part, the piano roll reviews read similarly to today's album reviews, but one of the most unique elements of "Player-Piano Notes" is the focus on performance and practice suggestions for player pianists. Each "Player-Piano Notes" column is packed with performance and practice advice on piano rolls of all types, from classical works to popular music and songs. Regarding a roll containing Smetana's *Bohemian Dance*, reviewer William Delasaire suggests: "The principal subject rushes upwards with a fine thrill, and provides an excellent study in rhythmical pedalling."[54] While many tips and suggestions are clearly specific to the technique required for the player piano, others focus on the development of musicianship that is not unique to a single instrument. "Played inattentively it is dull, but with a proper rhythmic accent and observance of dynamic contrast it not only provides good practice, but is an excellent example of early sonata form," writes Delasaire in another column.[55] The fact that this column continued successfully for more than half a decade—and in *The Musical Times*, no less, rather than a smaller journal that specialized in the player piano—suggests that there were in fact readers who were interested in guidance of this type. While it still had its critics in the 1910s and 1920s, the player piano had established itself as an instrument in its own right that offered users the opportunity to exercise their artistic imaginations and their manual dexterity while encountering new repertoire.

Firsthand accounts of users becoming acquainted with and practicing on their player instruments are rare, but one particularly detailed description of this experience was penned by Bertram Smith in 1911.[56] Smith purchased a piano-player for personal use in his drawing room and, according to his story, spent his first two days shouting at it in frustration. Smith's story narrates the way in which the instrument at first seemed to him incapable of producing finer gradations of expression, how he reached a point at which he was nearly prepared to throw it out, and how his struggle finally began to produce satisfying results. Smith concludes that after much practice, he found his player to be "an ally of wide and splendid capabilities."[57] Unlike some of the promotional literature for the piano-player, Smith does not make grandiose claims about the device or suggest that it enables him to perform at a level equal to the great pianists. His primary focus is on the way in which the piano-player democratizes classical music—a domain that was formerly inaccessible to amateurs without highly practiced skills.

Smith describes himself as a deeply committed music lover who is familiar with all of the Beethoven symphonies and would recognize any Wagner leitmotif, but who cannot play the piano himself. Writing of his frustrations with the difficulty of fully exploring the classical repertoire after ten years of commitment to "steady and consistent" attendance at concerts, Smith argues: "For if one really comes to examine it the world's music has always been remote, locked up from the generality of mankind. I have been more assiduous than most in my pursuit of it. I have attended literally hundreds of concerts. I have armed myself with miniature scores, and richly annotated them in red ink.... But what of the [Beethoven] Sonatas? I have heard perhaps ... a dozen of them.... With the best will in the world, and without missing any reasonable opportunity, we may well go to our graves without having explored one quarter of our heritage."[58] Smith was no casual attendee. And yet even when people such as himself had the opportunity to attend concerts regularly, they could not hope to apprehend all of the nuance and meaning in a work on just one hearing. But for an audience in the early twentieth century, one hearing may in fact have been all they would ever be afforded. Smith asks, "What, for instance, can the amateur who has had no special study hope to make of a Liszt Pianoforte concerto at a first hearing, and when may he hope to hear it a second time? Surely that great art-work had something more than that to tell him?"[59]

Smith is not arguing that the piano-player can replicate the full grandeur and timbral complexity of a concerto or symphony. Nor is he arguing that an amateur at a piano-player can produce a performance of a work on the level of a world-class artist. Smith openly admits its deficiencies. He notes that a trained pianist's skilled hands and practiced intellect can produce superior results, and that there is a significant amount of music that the player will never render in a truly pleasing manner. However, for owners who wanted an instrument with which they could enjoy music on their own terms, the piano-player was perfectly sufficient, and in some circumstances, perhaps even more desirable than the challenges and constraints of concert experiences. "The great pianists," Smith explains, "are not *here*: nor any pianist. I do not want to have to put on my dress-clothes and go and sit in a draught whenever I listen to music. I want to remain at home and take it at my leisure. And finally and above all there is the great fact of repertoire. With a few unplayable exceptions, all music is within my reach at last. It is *mine*, mine to explore and study and enjoy."

COMPLETING THE RECEPTION CYCLE

In these later trends in advertising, representations in popular sources, and instructional literature, we can see the concept of the player piano moving ever farther from its original identity as a simple machine, toward new uses as a tool for creativity, and an instrument in its own right. By charting this unfolding conversation, we can observe the way in which player devices emerged from their hazy beginnings as open-ended "new media," through a period of definition against a backdrop of earlier technologies, toward a unique and established purpose. During this time, a reversal took place in which the piano-player became a musical instrument, and its user became the piano player. Contrary to the anxieties of those who felt concern about mechanical music and its implications on the development of hard-won, traditional skills at the piano, the rise of the player piano did not signal the downfall of piano performance or spell the end of the profession of piano teaching.

I opened this chapter by relating the weak results of my efforts at the player piano and will close it by offering a contrasting story. Regarding my own encounters with the player piano, Sydney Grew and others would undoubtedly have criticized my ineffectual pedaling, my lack of pneumatic power management, and my failure to carefully study and plan accordingly to convey the structure of the work I was haphazardly playing. But it is one thing to claim that someone can master musicianship at the player piano with years of effort, and it is another thing altogether to see these claims borne out. Could years of practice really yield musically compelling performances?

In 2017 I attended a concert given by Rex Lawson, a musician who has spent decades giving professional performances around the world at his push-up Pianola piano-player. On the occasion I heard him play, Lawson's repertoire included classical solo piano repertoire, original works for player piano, orchestral transcriptions, and collaborative pieces with musicians on a range of other instruments. His recital opened with Chopin's Scherzo no. 2 in B-flat minor. In the realm of piano literature, few pieces could have matched this selection—with its tempest of wildly shifting tempos and dynamic contrasts—as an all-or-nothing wager on the piano-player as an artistic musical instrument and the performer as a true musician. The opening bars of the Scherzo begin sotto voce, pianissimo, followed by suspenseful silence and a blistering fortissimo. Lawson's performance swung back and forth between extremes of sound and stillness, full of tantalizing rubato and dramatic pauses. My gaze was fixed on his feet as he worked the pedals of the Pianola. Far from a mindless repetitive motion, his pedaling was nuanced, constantly shifting pace and strength to confer dynamic fluidity on the unfolding performance, always staying a step ahead of the next significant musical shift. Lawson hunched over his instrument, eyes fixed on the scrolling piano roll, reading its perforations as they passed, his fingers flicking back and forth to carefully manage the damper pedal and nuance his tempo. Tempo rubato is not cut into piano-player rolls, and neither are pedaling nor dynamics encoded in the punched paper. All of this musicianship was the performer's and his alone, and the Scherzo was the perfect vehicle to showcase the interplay of verve and sensitivity at his instrument. Lawson put the Pianola and the piano through its paces in an electrifying coda that drew wild applause from the audience at its conclusion. Despite the fact that I had studied this instrument extensively and came to this concert fully expecting Lawson's musicianship to be on full display at the Pianola, I was nevertheless spellbound and moved by the extent to which his artistry shone in this performance.[60]

Lawson opened this recital with a work that was chosen specifically to prove the piano-player's legitimacy as a musical instrument that requires a high degree of artistry and technique. However, if his entire program had been comprised of solo piano repertoire, it may have given the audience occasion to ask whether the Pianola wasn't merely some kind of "shortcut" to traditional piano playing—and this is the very type of accusation that is often leveled at many of the music technologies whose stories appear in this book. However, Lawson's program did not end there. Later in the evening, we were treated to performances of works written specifically for the player piano that showcased some of the different musical possibilities that the instrument could facilitate. Lawson performed works by Conlon Nancarrow, featuring studies that drilled deeply into contrapuntal rhythmic relationships that exceeded the ability of a human pianist to execute accurately. He also played transcriptions of works by Stravinsky, including a version of *Le sacre du printemps* that would have been entirely unplayable by a pianist with only ten fingers. The transcription took full advantage of the Pianola's dozens of fingers and superhuman speed to render on a single piano keyboard that which normally took a full orchestra to perform. These works conveyed a different idea: that player piano technology had more to offer than a streamlined way to play piano pieces. These instruments unlocked entirely new realms of creativity for composers, arrangers, and performers, and many of the works written specifically for the player piano suggest answers to the question of what is possible at the piano if artists are not limited by the human body.

Across a diverse program, Lawson and his Pianola demonstrated clearly that whether it was on stage in a concert hall or in the corner of someone's drawing room, player piano technology was more than a subpar stand-in for a pianist. On this instrument, a skilled player pianist could sensitively perform solo piano repertoire, a beginner could encounter and study music that may never have been programmed in a nearby concert hall, or composers such as Nancarrow and Stravinsky could make use of previously unimaginable technical capabilities. Music technologies like the player piano, when combined with the imagination and ingenuity of human users, create spaces for new forms of musical creativity. In this story, by tracing the trajectory of the player piano's meaning as it solidified, we can witness the shift from the 1898 slogan "this will play your piano" to the suggestion "with this, *you* can play your piano."

2. The Reproducing Piano

Capturing and Recreating "Living Presence"

On November 17, 1917, the New York Symphony Orchestra, conducted by Walter Damrosch, gave a performance of Saint-Saëns's G-minor Piano Concerto with pianist Harold Bauer. At the same time, nearly eight hundred miles away in Chicago, Harold Bauer also took another stage and performed a solo recital. This feat of duplication that seemed to feature Bauer in two places at once was made possible because Bauer's Chicago performance was delivered in person, while his New York appearance was recreated with a Duo-Art reproducing piano (see figure 7).

The New York Sun's account of the performance focuses on the reproducing piano's role in the concert: "When Mr. Damrosch stepped upon the conductor's platform to conduct the concerto he had a button pressed and the large concert grand on the stage fell into the opening bars of the concerto, and for thirty minutes, actuated by the Duo-Art record as made by Mr. Bauer himself, it played the work with the orchestra to the end."[1] Another account of the same performance, printed in the *New York Tribune*, painted a picture that focused less on the reproducing piano and more on Bauer's musicianship: "Mr. Harold Bauer, at the moment presenting a concert program in Chicago, a thousand miles away, had exhibited his highest art as literally as though he sat in person at the keyboard. His extraordinary genius transcribed upon a music-roll in the fullness of both its technique and its spirit was a present living actuality to every listener."[2] These two accounts frame the musical agency in this performance in remarkably different terms. In the summary from the *Sun*, the piano onstage at the concert simply "fell" into motion, and "it" played the concerto by itself. In the words of the *Tribune* write-up, by contrast, Bauer's musical genius "was a present living actuality to every listener," and the transcription of Bauer's performance onto a piano roll is mentioned almost as an afterthought. This

FIGURE 7. Sketch of the New York Symphony Orchestra concert that featured Harold Bauer's reproducing piano roll. *Source:* "A Notable Presentation of a Notable Instrument," *New York Tribune*, November 25, 1917, 6.

account conveys a sense of immediacy and human artistry, despite the fact that there was still no pianist present in the hall.

One review focused on artistry, while the other focused on the automation. One of these reviews was also an everyday reporter's write-up printed the morning after the concert alongside accounts of other musical performances in New York, while the other was in fact part of a full-page advertisement printed more than a week after the concert by Aeolian—the company behind the Duo-Art reproducing piano. Which was which? It seems reasonable to think that Aeolian would proudly feature its automatic piano and tout the way "it played the work with the orchestra to the end." However, the write-up that focused on the technology was not Aeolian's—its advertisement focused instead on Harold Bauer's pianism, the artistry of the performers involved in the concert, and the prominent musicians who were present in the audience, while the machine seems to be rendered almost invisible. But this was precisely the point. Reproducing piano companies like Aeolian were not marketing the idea of machine performance. They were working to create and fulfill a desire to faithfully capture and possess human performance and to market the belief that with enough measurements and data, the magic of a live performance could be revived at will.

This chapter traces a technological dialogue that extends beyond boundary-testing performances like Harold Bauer's and into the processes that reproducing piano companies used to create their products and the framing they used to sell them. By peeling away the veneer of companies' lofty technological and artistic claims, I show why reproducing piano rolls were not the scientifically objective snapshots that companies claimed they were, but were actually something even more interesting: a flexible and unique product of negotiations between pianists' musicianship, editors' technical knowledge and skills, the human desire for control, and the particular capabilities and limitations of the reproducing piano. The scientifically charged allure of reanimating a performer's aura extends back a full century before the motion-captured and painstakingly rendered performances of Maria Callas and Roy Orbison holograms were projected onto stages and accompanied by orchestras. Our desire to tame the ephemeral has a long history, but it is never as straightforward as we might like to believe.

PLAYER PIANOS AND REPRODUCING PIANOS: DIFFERENT INSTRUMENTS WITH DIFFERENT PURPOSES

Chapter 1 told the story of piano-players and player pianos, pedal-operated instruments that required a user's physical involvement and skill to make music. The reproducing piano, however, was exclusively an electrically powered device that delivered fully automatic playback from reproducing piano rolls without any input from its user. These instruments employed more sophisticated encoding technology in order to create accents, tempo changes, and dynamic gradations. They also arrived on the market later and cost far more money than their pedal-powered counterparts.

Some accounts of the history of the player piano erroneously situate the reproducing piano—with its more complex machinery, later development timeline, and higher price tag—as the culmination of player piano technology. Others conflate the piano-player, player piano, and reproducing piano by suggesting that all player piano technologies were created to reproduce the performances of great pianists. Still other writers freely mix and match the messaging that underpins advertisements or events featuring player pianos and reproducing pianos, suggesting that the ideas and motivations that pertained to one also pertained to the other. However, the reproducing piano was not simply a fancier player piano, and player pianos were not midway points in a chain of technological development that culminated in the reproducing piano. Describing them as such commits what Paul Duguid has called the error of technological supersession, in which newer and more complex technologies are thought to spell the end of simpler technologies.[3] The reproducing piano did not replace the player piano. Even at the height of their popularity in the United States, reproducing piano sales comprised only 10 percent of the market for player and reproducing pianos.[4]

Both instruments existed side by side for years, not as different price points for similar instruments, but as functionally different tools that appealed to different types of customers.

While both instruments used hole-punched rolls to store and play music, the artistic aims behind the reproducing piano and the pedal-powered player piano differed significantly. The player piano achieved much of its eventual popularity not as an automated playback device but as a democratizing instrument that provided its users with mechanical support at the keyboard while allowing musical control over many other aspects of the performance. The reproducing piano did not serve these same ends. The makers of reproducing pianos built instruments that they claimed could accurately and automatically play back the performance of a great pianist such as Harold Bauer, down to the finest detail. In short, then, whereas a player piano roll stored information that would facilitate future performances by a player pianist, a reproducing piano roll stored information pertaining to a past musical performance by a professional pianist.

These two types of instruments also diverged more sharply as they continued to develop. In the player piano's later years, advances in technology focused on the addition of new buttons and levers that were meant to provide users with new ways to shape their musical performances. At the same time, in laboratories dedicated to reproducing piano technology, researchers worked to invent increasingly precise techniques for capturing and recreating piano performances, and they marketed their devices on the basis of their machines' accuracy and sensitivity. This difference in purpose was clear enough that in 1917, a columnist published a piece discussing the future of both the pedal-powered player piano and its electric reproducing piano counterparts, describing them as "rival" instruments whose makers advocated for "two widely divergent views on the future of the player."[5]

The story of the player piano in chapter 1 revealed how a device that was marketed as an "automatic" instrument was not actually automatic at all, but a technology of mechanization that required continuous physical and musical labor to bring a unique performance to life. The reproducing piano, in contrast, does in fact deliver a fully automated performance from a roll of punched paper: flip a switch, and away it goes, keys bouncing up and down in front of a vacant piano bench, with nobody pumping pedals or shifting levers. The question is, whose performance is it? Is it a performance by the reproducing piano itself, as numerous concert reviewers suggested? Is it a performance by the original pianist, as the Aeolian advertisement and the bold lettering on the piano roll's box claimed? Just as Rousseau noted that Vaucanson's automaton flute player did not make music, but rather its mechanic made music, in the reproducing piano's history we can observe another partnership between creative humans and their automated tools. The process by which technicians encoded a performance onto a piano roll and a reproducing piano interpreted the data on that roll was not a straightforward matter of exact musical transmission. The "mechanics" in the story of the reproducing piano include a chain of pianists, technological innovators, and musical editors, much of whose work was invisible in the moment when the reproducing piano fell into motion, but all of whom contributed to the ongoing construction of the meaning of mechanically reproduced music in this time period. Though there are differences between the purposes and functions of the automation technologies explored in this book, one important element remains consistent across every story: the meaning of musical automation is less about the substitution of machine for human and far more about the negotiation of the relationship between the two.

ENDINGS AND BEGINNINGS OF THE REPRODUCING PIANO

The history of reproducing piano technology began with an indistinct murmur and ended with a crash—the stock market crash of October 1929, to be specific. The onset of the Great Depression that brought the Roaring Twenties to a grinding halt shrank the market share for luxury items such as the reproducing piano even smaller than it had previously been. Compounding the reproducing piano's difficulties during a time of economic hardship was the growing availability and quality of radio broadcasts and phonograph recordings. Reproducing piano and player piano companies were in competition with phonograph companies for more than two decades before their decline. Early advertisements highlight the versatility and affordability of the phonograph: "Why tie yourself down to the limited enjoyment that a piano or player-piano gives?" asks an advertisement for Victor Talking Machines. "The VICTOR is not one instrument, but every instrument. It delivers the whole realm of music into your hands."[6] At the bottom of this advertisement, and each one like it, the slogan "a 'Victor' for every purse" promises affordability to anyone interested in the machine, from the $500 model destined for the home of a wealthy owner, to the offer of "$10 for a serviceable one for the modest apartment."[7] Even the most expensive phonographs were significantly more affordable than reproducing pianos. Companies selling reproducing instruments could not compete on the basis of lower prices, and so they focused for many years on cultivating an image of luxury and exclusivity.

Reproducing piano roll catalogs from Ampico offer a glimpse of both the reproducing piano's image of luxury and its eventual decline. In 1916, the Ampico catalog was printed in an ornate book that generously spread 286 roll listings across seventy-two pages. After years of steady growth, the 1925 catalog was the largest and most lavish yet, quadrupling its length to a voluminous 351 pages, and bound in hard cover, with black, green, and gold detailing. The catalogs from 1933 to 1940 paint a clear picture of the difficult economic conditions of the Great Depression and the reproducing piano's diminishing market, with their smaller page sizes, their scant sixty-four-page length, and the addition of advertisements for accessories and cabinets. Similarly, the Ampico magazine, which was initially printed on high-quality paper, became the *Ampico Bulletin*, printed on a cramped and plain sheet of paper that included a list of discounted bargain rolls.[8] The writing was on the wall: in 1941, Ampico ceased roll production altogether.

The history of the reproducing piano's demise is fairly clear, but the details behind its rise are far murkier than those of the piano-player and player piano. One of the earliest attempts to create a system for capturing data from a real-time keyboard performance was a piano-harpsichord developed by Joseph Merlin in 1780. Merlin's instrument was equipped with a mechanically driven paper belt and sixty-one graphite pencils, which recorded the duration of each note depressed on the keyboard.[9] This innovative system lacked a means of playing back the pencil-marked record, however, and required a music technician to translate the graphite lines into a musical score, which a human musician could later play from, if desired.

It took more than a century before a complete system for capture, storage, and reproduction was developed, and it came from the small city of Freiburg, Germany, where a company owned by Michael Welte built orchestrions: automated music machines that used organ pipes and percussion to imitate the sound of an orchestra. The first reproducing piano mechanism on the market was the Welte-Mignon, and although we know its place of origin, accounts disagree on exactly when this device was originally available, with possible dates including 1904, 1905, and a vague guess at "about 1906."[10] More recently, Rex Lawson has pinpointed a location and date for the Welte-Mignon's first public showing, placing it at the 1904 Leipzig Autumn Trade Fair, which took place at the latest in August of that year.[11] The first Welte-Mignon device produced was a push-up cabinet-style model, although it was of a different sort than the push-up piano-players of the late nineteenth century, with its electric motor replacing the prominent pedals of the earlier devices. Across the Atlantic in the United States, The American Piano Company began selling its first model of Ampico reproducing pianos in November

1912, sixteen months before Aeolian introduced its Duo-Art in March 1914.[12] Each of these competing devices had in common a relatively similar system for reproducing performances from data encoded in a punched paper roll. However, more interesting than their similarities are the differences in how each company promoted its products, and how these efforts comprise a new framing of the relationship between humans and automation technology.

SELLING THE REPRODUCING PIANO:
STATUS AND "LIVING PRESENCE"

Although reproducing pianos were far more expensive purchases than traditional pianos, both instruments were still luxury items that required marketing strategies different from those used for everyday necessities. In order to convince potential customers to invest in a purchase as significant as a musical instrument, advertising worked to appeal to the emotions as well as the intellect. Companies put significant effort into painting a picture in the mind of the reader of the intangible benefits and cultural capital to which the instrument would open up access.[13] In advertisements for reproducing pianos, copywriters sought to conjure images in readers' minds of status, emotional satisfaction, and possession of the ephemeral. From the reproducing piano's early years until approximately the middle of the 1920s, advertisements gave relatively little attention to the details of the automation technology inside the instrument and focused instead on selling a feeling of exclusivity and cultural edification that could elevate the status of wealthy buyers. Many ads were featured in publications with an affluent readership and shared advertising space with other luxury products. In a 1912 supplement to *Country Life*, an advertisement for the Welte-Mignon shares a page with an advertisement for freezing rooms in which one could store one's collection of furs.[14] Elsewhere, an ad for the Ampico is positioned directly above an ad for chauffeurs' uniforms.[15] One full-page advertisement for Ampico reproducing pianos that was printed repeatedly in the *St. Louis Post* uses ornate lettering alongside emotionally charged descriptions to present one of the most extraordinarily sensationalized advertisements to ever promote a reproducing piano (see figure 8).

The leading line on the advertisement—which ran shortly before Christmas—simply suggests in elegant script that the Ampico reproducing piano is "The Finest Gift in all the World!"[16] This heading and sketch are accompanied by a single sentence, which does nothing to describe the Ampico—nor does it even mention the instrument: "That superlative thrill, that moment of pure conquest which is vouchsafed but seldom in a

FIGURE 8. Excerpt from an extravagant Christmastime advertisement selling the Ampico reproducing piano. *Source:* Conroy's, *St. Louis Post – Dispatch*, December 11, 1921, B17.

lifetime—and for which a whole lifetime of effort is not too dear a price to pay—is yours, sir, when you say to her Christmas Morning, 'Yours, My Dear!'" This caption is not describing a piece of technology or a musical instrument; it is selling a feeling of exhilarating accomplishment. In 1920 Ampico grand pianos were priced from $2,000 upward, and in 1921 a top-of-the-line model sold for $4,000.[17] Given that the average annual salary for full-time employees in the United States at this time was $1,283, saving up for this "superlative thrill" may well have taken nearly a "whole lifetime of effort" for an average worker.[18]

Farther down in the small-print body of the advertisement, Ampico makes increasingly dramatic claims: "The AMPICO for yourself—the AMPICO for all the family, it means a new lease on life, a new interest in living."[19] At this point, one has to wonder a little at these copywriters' estimations of either the psychological condition of their customers or perhaps their credulity, but the ad continues to escalate: "Joy, joy, that you never dreamed existed is yours when the AMPICO comes." Unimaginable joy? A new reason to keep living? The claims in this advertisement are patently outrageous, promising buyers that the purchase of a reproducing piano could confer things that no musical instrument company could ever sensibly promise. But while this advertisement might be good for an incredulous laugh, it also throws a spotlight on the new concept of automation technology that is crystallizing in the story of the reproducing piano. Compared with the player piano—whose early advertising copywriters focused on its labor-saving potential, aligning it with household appliances and sewing machines—reproducing piano advertisements like this one were not selling the idea of automation as a support for challenging or tiresome tasks. They were selling the idea of automation as a technology that could impart control over unmasterable forces, possession of the ephemeral, and the attainment of previously exclusive echelons of privilege.

Most (or perhaps even all!) other reproducing piano advertisements took a more subtle approach than this Ampico advertisement did, but many of them contained similar themes of pleasure, control, and status. A large number of ads include photographs of the drawing rooms of wealthy families who owned these instruments and supply lists of the names of prominent purchasers (including political figures, artists, and business owners, such as the Prince of Wales, the president of Cuba, and Mrs. Alfred G. Vanderbilt).[20] These ads featured testimonials describing how reproducing pianos added enjoyment and satisfaction to their purchasers' lives. One such advertisement ran in the *New York Tribune* in 1918 and features the author Dr. Cyrus Townsend Brady. The ad contains scarcely any advertising copy about the reproducing piano itself. A photo of Brady seated next to his reproducing grand piano occupies a third of the advertisement, and another half of the page contains an uninterrupted quote from a letter by the famed writer, which begins by weaving a story in the second person:

> You come home from a day's work thoroughly tired out. You find the evening papers flat, stale and unprofitable; the war news adds to your depression. What can relieve it so well as a little music? The great Godowsky is playing to-night at Carnegie Hall, or the fiery and passionate Volavy is to be heard in recital at a nearby theatre; and there are other exponents of the great art available. But you feel strangely disinclined to don your evening clothes and adventure forth in cold or rain. Pipe and slippers, easy chair and fire, win the day, or the night, rather. And then you reflect that you are not going to lose anything after all. For while you take your ease in your house Mr. Godowsky and Miss Volavy and all the rest play for you exclusively, and for you alone.[21]

Brady's disinclination to leave home for a concert hall echoes that of Bertram Smith—who wrote about the advantages of the pedal-powered piano-player seven years earlier—down to the same complaint of having to don one's dress clothes in order to hear good music: "The great pianists are not *here*:

nor any pianist," Smith wrote in 1911. "I do not want to have to put on my dress-clothes and go and sit in a draught whenever I listen to music. I want to remain at home and take it at my leisure."[22] Beyond this point, the similarities begin to dissipate. Smith, the proud owner of a pedal-powered instrument, would have seated himself before his piano-player and physically pedaled his way with thoughtful effort through the pieces he wished to enjoy. Easy chairs and pipes were not a part of his evening experience. For Smith and his piano-player, there were still no pianists present in the room, nor did the makers ever claim there were. But with the reproducing piano, companies spoke of how great pianists were present in the room in some intangible way, playing privately for owners' pleasure. Those who owned a reproducing piano did not need to exert physical labor for their music. In an intriguing reversal, then, the device that saved the most labor was not actually the piano-player or the player piano—the ones that were marketed as labor-saving devices—but the reproducing piano.

Not only could reproducing pianos' owners opt out of sweating and straining their way through a Beethoven sonata, but as Brady's testimonial suggests, they could feel confidence that the concert pianist's roll that was effortlessly played back was also delivering a far superior interpretation of that sonata than any amateur could manage, whether on a player piano or a traditional piano. Regardless of what any of these companies and users were claiming, however, it is clear that there were no pianists, great or otherwise, truly present in the drawing rooms where these instruments stood. However, references to an intangible sense of artistic presence were another significant theme across reproducing piano advertisements. Many ads sought to convey the idea that an artist's concert presence could be fully recreated by the reproducing piano. These advertisements draw up just barely short of saying that the performer is bodily present when their rolls are played back, and instead hint that some immaterial but significant element of the professional recording artist's aura can be conjured in one's living room through the reproducing piano. Somewhere in between the ghostly dance of the piano keys and the vacant space on the bench, a sense of invisible artistic intention seems to drive the music being advertised here. Perhaps, they imply, the pianists whose names adorn these rolls are not physically present, but neither are they fully absent.

Each reproducing piano company focused on a slightly different aspect of this almost-presence, but they had in common the idea that this presence was an active, or even somehow "living," human presence. In Brady's account, while he sits in his easy chair, he is entertained not by an artificial simulation of a performance, or even a playback of a performance, but by the full experience of "Mr. Godowsky and Miss Volavy... play[ing] for you exclusively." The reproducing piano is framed not just as an automated player piano, but an opportunity to access something exclusive. This type of presence is similar to the advertisement recounting Harold Bauer's reproducing piano concerto, in which the pianist's "extraordinary genius... was a present living actuality to every listener."[23] Welte-Mignon also took up this same approach and edged even closer to making metaphysical claims about "living" presence in a 1911 advertisement: "The Welte-Mignon Autograph Piano Is The Living Soul of the Artist." The body of the ad further emphasizes, "It has the *living touch* which differentiates the Welte-Mignon from every other piano playing device in the World. So accurately does the Welte reproduce each note—each delicate shade of contrast—and the individuality of the player, that famous critics declare it impossible to believe that the artist in person is not actually playing."[24] Brady's testimonial for the Ampico also draws on this concept of living presence, but situates the reproducing piano owner's control and possession of quality performances in the context of the loss of an artist who has passed away: "And then, one day, you read that some favorite artiste like Teresa Carreno is dead. Her hands will witch no more noble music from the keys. But she being dead yet plays on. The frail records give her immortality. You need not long for the magic touch of vanished hands. It is there. You have it, at your command. For the Ampico, which gives permanence to the evanescent, which embodies the immaterial, is an immortal instrument."[25] The ghostliness of a reproducing piano performance was amplified even more when the "vanished hands" were not simply in a different location on the globe, but seemed to be resurrected from beyond the grave to invisibly grace the keys of the reproducing piano, offering a performance that was at once evanescent and immortal.

The historical importance of advertisements and testimonials like these lies less in assessments of the truth of their claims or in estimations of how they might have impacted sales, than in how they connect to the broader landscape of debates around the purpose and meaning of automation technology in music. Both the Welte-Mignon advertisements and Brady's comments draw on rhetoric that also appears prominently in promotional material for the radio and the phonograph, which were said to have the power to "suppress absence" or even "conjure the dead."[26] As John Durham Peters has shown, advertisements for telephones, phonographs, and radios worked to "reassure their customers by reconnecting the mechanically reproduced representations to an originating body (via testimony and authentication)."[27] Without this authenticating link, phonographs and radios were merely machines making disembodied sounds. Similarly, reproducing piano companies' advertisements and testimonials focused carefully on painting a picture of this automation technology not as a machine-like rendering of unfeeling noises, but as a gateway to a personal encounter with a human musician whose work was faithfully and transparently reproduced for an appreciative listener. Reproducing piano companies' reassurances regarding the carefully preserved connection between the "originating body" and the playback included musicians' signatures printed on their rolls, artists' testimonials published in brochures and advertisements, and photographs of the pianists that were taken at their reproducing piano recording sessions. These tangible records of human presence were touted in order to anchor the reproducing piano's potentially eerie performances to a moment when living humans were actively involved in the music-making process.

In the twenty-first century, many listeners have accepted as fact these companies' claims that reproducing piano technology was neutral and transparent, believing that they offer us a window through which we can witness without alteration the playing of great pianists of the twentieth century. Many writers—even academic writers—have accepted the story of automation technology told by companies like Ampico, who assured us that with enough data, with sufficiently fine-grained measurements, and with even more precise mechanisms, automated machines could fully capture and recreate human musical performance. It's an alluring promise—the idea that something as ephemeral as the experience of witnessing a world-class musician in concert can be captured and preserved and replayed indefinitely without the loss of that intangible feeling of "living presence." But unfortunately, this simply is not the case. Automation technology is not neutral. Its parameters and definitions are culturally constructed, built to align with the values of its creators in the context of complex human desires. The reproducing piano was no exception. Beneath the marketing tactics and testimonials that suggested an unbroken link between pianist and listener, the so-called automatic processes that companies claimed could objectively capture and recreate a performance required a great deal of subjective human involvement. The efforts of the "mechanics" in the story of the reproducing piano, however, did not fit the narrative of scientific objectivity, and were therefore carefully concealed.

INSIDE THE "MARVELOUS MECHANISMS": WELTE'S AND AMPICO'S EARLY YEARS

Many reproducing pianos still survive in good condition, but the recording pianos—the instruments on which companies captured the performances of the great pianists of the day—have been dismantled or destroyed, and many of their secrets perished with their inventors as closely guarded proprietary knowledge. Technical descriptions and firsthand accounts of the recording process frequently conflict. Details from reproducing piano companies themselves were often deliberately vague, especially when describing the way in which dynamics were reproduced.

Among the three largest reproducing piano companies, Welte-Mignon was the most secretive about its recording processes, and avoided disclosing them altogether, describing their system in obtusely enigmatic terms across nearly twenty years of ads. In 1909, a Welte-Mignon advertisement in the *Los Angeles Times* told readers, "By another secret process these performances are reproduced, always retaining every characteristic of tone, touch, phrasing and volume—just as expressed in the original performance."[28] Given the importance of these sizeable but vague claims, it may seem surprising that Welte-Mignon expected potential customers to take them at face value. In 1909, however, the company did not yet have any serious competition on the American market with which its products could be compared, and this lack of detailed information about "secret processes" lent its instruments a certain mystique.[29] Two decades later, however, even with newer competitors on the market, Welte-Mignon continued to maintain silence regarding its methods. One ad claimed, "Through this marvelous mechanism the artist's playing is actually photographed—by a secret process known only to Welte-Mignon. That is why these are the only recordings that reproduce every delicate shading—and why editing a recording is not necessary with Welte-Mignon."[30] The "marvelous mechanism" touted here still receives no explanation in the ad, even though this advertisement appeared nearly twenty years later than the 1909 ad that used the same "secret process" phrase. Notably, Welte's claim to authenticity here includes the boast that its records do not need to be edited—in other words, according to Welte, its automated process provides an unaltered glimpse of the pianist's playing.

Welte's claims that its methods "actually photographed" the playing of its performers seems to be a stretch bordering on a falsehood. No accounts from artists recording for Welte-Mignon indicate that there was any type of performance capture taking place that paralleled photography in any meaningful way, and the rough information available concerning this company's technology suggests the use of methods that not only were distinctly nonphotographic, but also would have been unable to fully recreate expressive shading to the extent promised in advertisements. Welte's secrecy here is significant not only as part of a marketing strategy, but also because of the way in which it creates a "black box" around the technological processes used in the creation of its rolls. Welte claims to employ a process that faithfully mediates between an original performance and its reproduction, and customers are asked to believe these claims without knowledge of the technology's inner workings.

The concealment of these capture processes did not prevent writers from speculating about the possibilities. A description from the liner notes of a 1963 album titled *The Welte Legacy of Piano Treasures* suggested that the Welte-Mignon recording process involved a set of carbon rods attached to the keys of the keyboard, explaining how pressing piano keys would cause these rods to dip into a trough filled with mercury.[31] The contact between the rod and the mercury would produce a current, which in turn caused an ink-coated, wedge-shaped rubber wheel to press with varying force against a moving roll of paper as the pianist played. This roll would then be punched to translate the inked performance record into perforations playable on a reproducing piano in a customer's home. Kent A. Holliday took issue with this carbon-and-mercury theory, noting critical mechanical issues with a system of this type, and also pointing out that surviving master rolls from Welte disprove the possibility of this method having been used.[32] In a recollection that casts further doubt on the veracity of the technical elements of the carbon rod method, Ken Caswell spoke with *Welte Legacy* production assistant Ben Hall, asking whether the descriptions of the Welte recording process were true to reality. Hall's brief response is revealing: "Well," he said, "we had to sell the recordings somehow!"[33] Recent work by player pianist Rex Lawson draws together three other accounts of the Welte recording process and makes a strong case that explanations including the carbon rod theory are the unclear and contradictory efforts of individuals who lacked an understanding of basic piano mechanics and, furthermore, had likely never actually seen the Welte recording mechanisms.[34]

It is improbable that we will ever receive a definitive answer regarding the exact details of the "secret process" behind Welte's recording process. As far as we know, no recording piano system survives intact today, leaving us to piece together our best estimates based on incomplete recollections, master rolls, and a handful of fuzzy photographs. As fascinating as it could be to gain concrete answers to these questions, however, attempting to track down material information in order to validate Welte's claims is a wild goose chase that would likely reveal not an elegantly automated solution, but a partially automated system over top of which creative humans made music. Welte's intentional obfuscation and misdirection suggests that the answers do not lie with the machinery, but with the relationship that was constructed between the machines and the team of artists and technicians that worked with them. However, the stakes for the mystery surrounding this recording process are higher than merely setting the record straight. In any story of a contested automation technology in music, half-truths and myths are often just as important as the facts because of what they reveal about the values of the storytellers. The important attribute that all of the illusory accounts of the secret Welte recording process have in common is the absence of human intervention. Like Vaucanson's duck, which was "partly fraudulent and partly genuine . . . partly transparent and partly ingeniously opaque," companies were open about certain aspects of the recording process while actively concealing others, and although recording pianos did capture certain data from the pianist's original playing, expert human intervention was required to transform that data into a reproducing piano roll that could deliver a satisfactory musical performance. A great deal of the living touch was, in fact, that of skilled editors, not a direct transcription and reproduction of the artist's playing. The history of the Ampico's proprietary reproducing systems, which were explained in far more detailed than Welte's, sheds helpful light on the tension between technical data and artistic sensibility.

The American Piano Company's Ampico Re-Enacting Piano was the first reproducing piano to be built in the United States, arriving on the market in 1911, more than half a decade later than the Welte-Mignon. The inventor behind the Ampico's early success was a former employee of the American Pneumatic Service Company named Charles Fuller Stoddard, who quit his job in order to devote a year and a half to the private development of a system of musical dynamics production for automatic pianos.[35] Stoddard brought in a stream of piano builders to view his system, in the hope that someone would be interested in purchasing it. Although he received little interest at first, the American Piano Company eventually committed to buying the invention, and the company made Stoddard the head of its player piano department, which sold the Ampico under the name "Stoddard-Ampico" for several years.[36]

Far more information survives about Ampico systems from patents and interviews than about the mystery-shrouded Welte-Mignon. Stoddard applied for no less than twenty patents with the US Patent Office, detailing the workings of his systems for recording and reproduction, which leaves very little to the realm of "secret processes." Stoddard's early reproducing piano model had eight levels of dynamic intensity—though, significantly, there were not yet any systems to handle crescendos or diminuendos.[37] His system for recording dynamics employed electric contacts placed on the piano keys, which would create marks on a moving sheet of paper. When a pianist struck a key to play a loud note, the length of time required for the key to reach the bottom of the key bed was less than if the pianist pressed the key slowly to obtain a soft sound. Stoddard's system recorded this span of time with longer and shorter marks on the moving paper.[38] This process seems far more objective than the mysterious Welte-Mignon methods, but one of the most surprising twists in Ampico's technological history is that, in spite of all of Stoddard's innovations, the company appears to have made little if any use of them at all for fifteen years—fully half of the company's existence.

Although Stoddard's patents tell the story of his data-gathering breakthroughs, historical accounts and interviews with Ampico staff drawn together by Larry Givens suggest that prior to 1926, no dynamics whatsoever were recorded in Ampico's capture process.[39] Despite having the ability to capture detailed velocity data, dynamics—according to a range of sources—were added by hand, and not by the pianist but by an editor. Lawson confirms Givens's conclusions, drawing together descriptions from Ampico editors Angelico Valerio and Edgar Fairchild, neither of whom has any recollection of functional dynamics recording systems prior to 1926. Valerio, who began working for Ampico in 1923, describes the dynamics editing process at the time he was hired: "We'd know generally what dynamics to put in, because any piece they played we would have the music for it. We would read it over ourselves if we didn't know it, and we'd get a general idea of what they wanted."[40] In the same interview, Valerio was asked how the editors made use of Stoddard's dynamics data, but Valerio insisted that "there were no dynamic markings" prior to 1926.[41] Fairchild corroborates this recollection, noting that Stoddard's mechanism "recorded the notes, the sustaining and una corda pedalling, and nothing else."[42] A studio master at Ampico, who chose to remain anonymous, revealed that the goal was merely this: to convince not only potential customers, but also the recording pianists themselves, that every detail of their dynamics and technique was being captured objectively.[43]

We know for certain that Stoddard did in fact design a dynamics recording system in 1908 and applied for patents for it, but either this system was not used when recording Ampico rolls, or else the rough lines marked by the electrical contacts were simply overlooked by editors who chose to rely instead on their ears, their scores, and their musical instincts. The anonymous Ampico studio master admitted that occasionally, Ampico staff would use a phonograph record to do a test run before a pianist arrived to record, marking the wavy line on the score based on what they could hear on a phonograph recording by the same artist.[44] Ampico's editor in chief reported that occasionally the staff would obtain a first test run live from the pianists themselves in the Ampico recording room under the pretense of waiting for the equipment to be ready to use.[45] During this test run, Milton Suskind, who served as editor at Ampico from 1917 to 1925, would record data on the dynamics by means of a system that was known as the "Cookie Chronograph":

> On some such pretext as "timing the performance", or "killing time while they get the equipment ready", he would call for a complete run-through. During this performance the 'Cookie Chronograph' (Suskind himself) would "record" the crescendo pattern by drawing a continuous line on the composition's music sheet—the bottom of the bass staff representing pianissimo; the top of the treble staff representing fortissimo. . . . During the actual recorded performance, Cookie would again follow on the music sheet, this time marking the accents above the treble staff. A short line directly above the note indicated a soft accent only slightly above the basic volume; a long line denoted a heavy accent. That was the "Cookie Chronograph."[46]

Lawson suggests that highly practiced editors like Suskind simply would have been able to do a better job of shaping a convincing portrayal of a performance's overall sound than could have been obtained by rigorously following the data from Stoddard's early dynamics recorder.[47] This is because Suskind and his colleagues were intimately familiar with the nuances and quirks of the reproducing piano's dynamic steps and crescendo functions, and furthermore, had an intuitive feel for the time required for the reproducing piano's pneumatic systems to accomplish large dynamic changes. Often the best musical decisions that took the reproducing piano's affordances into consideration would not have corresponded directly with the data. In other words, the final product was far less an automatically and objectively rendered snapshot of an original performance and far more a carefully curated collaboration between a pianist, a team of editors, and a machine's particular pneumatic capabilities and limitations. Even though the reproducing piano's playback function was unchanging and automatic, the process of creating the roll was intuitive and nuanced. However, rather than embracing this flexible approach that drew on the knowledge and musical abilities of the team, Ampico treated the human influence in its process as a problem to be solved and its quest for objective automation processes as a publicity opportunity.

THE SPARK CHRONOGRAPH: SELLING OBJECTIVITY AND FIDELITY

In 1926, Ampico announced the invention of an automated mechanism that could capture data from the recording piano with unprecedented accuracy. This system, named the spark chronograph, was one of the most thoroughly documented and publicized technological advances in reproducing piano history, and the fanfare around its introduction further solidified Ampico's approach to scientific measurement as a proxy for performance fidelity. Ampico had long employed marketing strategies that focused on themes of technological precision, automation, and high-quality reproduction, and so the development of the spark chronograph provided the company with an opportunity to publicize photos that supported its image by documenting research and development processes taking place in its facilities. These photos appeared in advertisements, journals, and even in service manuals for Ampico instruments.

A collection of laboratory and recording studio photos included in the 1929 Ampico *Service Manual* shows several men in lab coats concentrating on an array of precision tasks in a series of clean and sparsely furnished rooms lined with shelves and workbenches (see figure 9).[48] Whether the lab coats were just for the publicity photos or Ampico researchers bustled around in them at all times, we can only guess, but the setup in these photos, from the pristine lab coats to the unidentified machinery with which the technicians are tinkering, seems tailored to convey a sense of scientific precision. Central to Ampico's ad campaigns was the idea that the essence of a performance could be precisely captured and recreated, if sufficient detail were obtained through scientific measurements. Ampico sought to position the reproducing piano as a technology that could objectively break down the actions of the human body at the piano into a series of data points and then reconstitute those data points into a high-fidelity reproduction. While Stoddard had already been working toward these goals for the better part of two decades, the spark chronograph was the tool with which Ampico sought to cement an image of a scientific approach to reproduction in the minds of the public.

The most famous fruit of the Ampico research lab's efforts, the spark chronograph (see figure 10) was introduced in early 1926, two years after Stoddard had hired physicist Clarence Hickman to join his team.[49] Hickman published a paper in *The Journal of the Acoustical Society of America* in 1929, in which he detailed the inner workings of the spark chronograph.[50] The chronograph used a lightweight silver contact attached to the shank of

Woodworking Department of the Research Laboratory.

FIGURE 9. Photo of the woodworking department of the Ampico research laboratory. Courtesy Edgerton Mechanical Music Library, Morris Museum, Morristown, New Jersey, USA.

each hammer on the piano. When a pianist pressed a piano key and the hammer moved toward the piano strings, this electric contact passed and touched two fine wires in sequence. The chronograph used a spark to punch a pair of holes in a moving roll of paper, creating a visual record of the amount of time that elapsed between the silver contact touching the first and second wires. The distance between these holes could then be used to calculate the exact velocity of the piano hammer for every single note.[51]

Hickman's article summarizes the detailed experimentation that researchers undertook in the Ampico laboratory to determine the best materials to use, the accuracy of piano hammer velocities given constant pneumatic pressure in the Ampico, and the optimal distance between the two contact points.[52] The significance of the spark chronograph had less to do with any audible differences in the piano rolls themselves and much more to do with the importance of the public discourse about this technological development. Reproducing piano companies had spent more than two decades working to sell the idea that reproducing pianos played back professional performances in accurate detail, but descriptions of how this was accomplished were problematically vague, and methods were shrouded in mysterious terms such as "secret process" and misleading descriptions like "actually photographed."

One of the most interesting instruments used by the Research Laboratory in the development of this new Ampico is the chronograph.

This instrument measures hammer velocities so accurately that it divulges differences in loudness as small as one one-hundredth of what the ear can detect.

FIGURE 10. Photo of Ampico's chronograph, similar in appearance to promotional photos of the chronograph that appeared in other publications. Courtesy Edgerton Mechanical Music Library, Morris Museum, Morristown, New Jersey, USA.

With the spark chronograph, at last, Ampico could put its technology on display in order to convey a sense of transparency and scientific rigor to the public. The highly publicized details of this mechanism were offered up as proof of Ampico's claims to precision. Patents and academic journal articles were important outlets for this narrative, with these publications conferring a sense of serious scientific rigor on the recording and reproduction process. Shortly after the spark chronograph's introduction, Ampico also turned to trade journals to bring its message to new audiences of piano technicians.

"The Microscope of the ear," ran a heading on a full-page advertisement by the American Piano Company in *The Tuners Journal* in 1929.[53] This advertisement introduced the spark chronograph, describing it to piano tuners as "an instrument which makes the finest split-second watch ever built take a back seat." Beneath a photo of what is perhaps the chronograph itself, or some part of it, the ad works to flatter the readership and to link the chronograph to other scientific tools: "The ordinary microscope reveals marvels imperceptible to the unaided eye. The Chronograph does the same for the ear. The well-trained, sensitive ear of the tuner hears differences in sounds that are not apparent to the ordinary listener. Yet, were his ear 100 times more sensitive, this instrument could measure hammer blows more accurately than he could determine differences in the resulting sounds."[54]

Here, Ampico takes pains to compliment the tuners themselves before touting the company's own technology, connecting the sensitivity and accuracy of the chronograph to that of the tuners' own hearing capabilities. Compared with Hickman's paper in *The Journal of the Acoustical Society of America*, however, this depiction of the spark chronograph offers almost no explanation as to how the machine works. Even the photograph itself, which occupies nearly half of the ad, is unlabeled and offers no assistance in comprehending the chronograph. A thin booklet of paper lies open on a table, in front of a dark case and shelves filled with dull wooden and glossy glass components. A pair of round gauges suggest the ability to measure something, but the reader knows neither what nor how. The purpose of this photograph—and of the ad as a whole—has little to do with improving readers' understanding of the spark chronograph, and more to do with conveying a sense of complexity and elite precision.

Ampico pointedly focused on this scientific approach to recording in order to make the claim that the reproduction process prioritized objectivity and fidelity and could therefore provide completely faithful access to the original performances. This seemingly straightforward and neutral approach to measurement and reproduction, however, was loaded with tacit assumptions about piano technique, performance, acoustics, and the definition of musical fidelity. Even Stoddard's research relies on a lengthy history of thought about the relationship between music and the mechanics of motion, which Nick Seaver has traced back to certain schools of nineteenth-century German piano pedagogy—which treated the acquisition of piano technique as purely mechanical—and Hermann von Helmholtz's textbook, *On the Sensations of Tone as a Physiological Basis for the Theory of Music*.[55] In this view, the piano keys were an interface that received physical input from the pianist's hands and translated it into hammer-strikes on piano strings. If those hammer-strikes could be precisely measured and recorded, and hammers could then be programmed to strike the strings with exactly the same speed and timing, then, Ampico argued, the reproducing piano delivered not just an imitation but the actual playing of the original pianist. Capturing and "making a performance happen again" was just a matter of automating an identical sequence of events.[56] By eliminating the subjective assessments in processes like "the Cookie Chronograph" or of the editors who made rough drafts of dynamics based on phonograph recordings, Ampico claimed that it was giving its customers unmediated access to the work of the great pianists.

Hammer velocity, however, was just one possible approach to the measurement of piano technique. By choosing to focus on this single aspect of the piano action, the company was also choosing to overlook many other measurable phenomena: the vibration of the strings, the acoustics in the room, and the movements of the performer's fingers, to name a few. By choosing a variable to measure and building up an image of objective scientific precision, Ampico worked to establish and defend its own subjective definition of fidelity, which in turn mediated the relationship between human pianists and the reproducing piano's performances. As Nick Seaver notes, "Rhetorics of fidelity are not just technical or commercial details, but meaningful interpretations of the relationships between human and technological identities."[57] The same is true of rhetorics of automation. What specific actions are automated? How, by what means, and why? Is an automation technology a support for human labor or a replacement for it? Does it kill creativity or augment it? The stories we tell about the purpose and function of automation technologies often hold more power than the technologies themselves, because any technological narrative functions only to the extent to which people are willing to believe in it. The reproducing piano sold units and rolls based on the belief that its automated processes could do what they claimed: reproduce the playing of the great pianists. Ampico's scientific approach to performance capture through the measurement of hammer velocity successfully did one thing: measured piano hammer velocity. The leap from these data points to a belief in the fidelity of a reproducing piano roll, however, was not scientific, but rhetorical.

Beyond the subjective relationship between hammer velocity and human performance, the reproducing piano posed one further unsolvable problem for technicians and artists: no matter how much detailed information was obtained from the spark chronograph, and no matter how precisely a technician measured the hammer velocities for every single note in a performance, the reproducing piano itself was still functionally incapable of playing back more than two levels of dynamics at any one time. Furthermore, the pneumatic systems of the reproducing piano simply could not keep pace with every split-second dynamics shift in a human pianist's playing. No amount of technological precision on the recording end could circumvent these basic limitations on the reproducing end, and even the most responsive systems still relied on editorial compromises in order to work within the limitations of the reproducing piano.

In order to understand what these limitations meant in practice, consider the challenge human editors faced when presented with an enormous record containing the spark chronograph data for thousands of notes, each one with its own measurement, and each one slightly or significantly different than the notes that neighbored it. How should that data be objectively converted into any kind of overall dynamic level applied across half the piano, that was both objectively faithful to the numerical data and musically coherent? Should the editor simply take the average of all of the dynamic levels of the notes at a given point in time? Tweak the numbers a little bit in order to preserve the effect of a melodic line? Compromise in order to balance a chord that had more notes in the bass than the treble? Considering these challenges, it may seem almost impossible for any technician to hope to render a pianist's performance accurately and musically. And yet not only do the rolls we have of these pianists' playing sound nuanced, sensitive, and musical, but the pianists who listened to and commented on their rolls did sign off on them, saying they were satisfied with their records. The editors of these rolls were highly skilled at working with the resources they had available to them: pneumatics, data, and their own ears and musicianship. Experts at manipulating the systems they were familiar with, these professionals developed techniques for working around the challenges inherent to their proprietary system, and the result is a characteristic sound that is recognizable across a given company's rolls.[58]

In spite of what the reproducing piano companies worked relentlessly to portray to the public, these stories of technological challenges and editorial prowess show that reproducing piano rolls were neither objectively automatic reproductions nor acoustic "photographs." One description that appears frequently across reproducing piano advertisements, testimonials, and literature refers to the reproducing roll as a "portrait" of an artist's playing. This term suits the process and the product much more accurately than "photograph" or even "reproduction" does. When someone sits to have their portrait painted, the artist seeks to render important aspects of their appearance, representing a living, breathing subject onto a two-dimensional canvas. Working with the limitations of pigments and brushes and paints, the artist must account for variation across repeated sittings, use their discretion to make the portrait's subject appear more attractive, and downplay any less-appealing features. In a similar way, roll editors would—sometimes with a performer's supervision—make edits to correct wrong notes and enhance

74 / Chapter 2

or clarify certain passages' dynamics or articulation. Ultimately, however, a painted portrait is not an exact duplicate of its subject in any meaningful way, and neither is a reproducing piano roll. A roll is a unique artistic perspective on a piece of music that worked with certain material and acoustic parameters. It is a collaboration among pianists, editors, and machines that were built and marketed as an opportunity to encounter human musicianship, and this opportunity for an encounter or an experience is how pianists themselves often spoke about their reproducing piano rolls.

PIANISTS RECOMMEND THE REPRODUCING PIANO

In spite of the limitations of reproducing pianos' systems, piano rolls could be crafted and edited to reflect pianists' individual styles well enough that the musicians were willing to attach their names and reputations to the rolls they recorded. Many of these early twentieth-century pianists gave testimonials and interviews, which provide insights into these artists' attitudes toward this technology. Testimonials were a favorite advertising tool for reproducing piano companies. When artists recorded for a particular system, their testimonials usually appeared in newspapers and magazines shortly afterward.[59] Companies collected large catalogs of these endorsements from pianists, composers, and conductors, and the themes that consistently weave through the comments and responses include a range of remarks on fidelity of reproduction and the educational value of the reproducing piano. These professional musicians—including Percy Grainger and Ernest Schelling, whose interviews are examined later—chose their words carefully, and in doing so, they reveal that they may not have had a naïve or unrealistic view of the relationship between their performances and the reproducing piano rolls that bore their names.

"The Duo-Art brings the playing of Percy Grainger to your home as realistically as if he were seated at your piano," reads the caption of an advertisement printed in *The New York Times* in 1917, above which appears a sketch of the Australian pianist and composer playing for a group of people seated casually in a domestic space (see figure 11).[60] A transcript of an interview conducted by Aeolian with Percy Grainger comprises the majority of this advertisement. Much of the interview features Grainger's unreserved praise for the Duo-Art's ability to provide people with more opportunities to listen to music and for its flexibility as a home instrument that can serve as a hand-played piano or a playback device. The pianist also suggests that the Duo-Art has given him the opportunity to hear what his playing would sound like from an audience's perspective. In a particularly revealing section

FIGURE 11. Image from a promotional interview with Percy Grainger. *Source:* The Aeolian Company, Advertisement, *New York Times*, May 23, 1917, 7.

of this interview, Grainger's carefully chosen words shed light on his views of the strengths and weaknesses of the reproducing piano:

> [GRAINGER:] "When I am *en tour*, my mother may have [the Duo-Art] to reproduce my records, and, for the time, I am with her in spirit—the Duo-Art reproductions are so vividly like my playing."
>
> "Altogether, the Duo-Art is quite wonderful indeed—one of the greatest marvels I have found in your remarkable America."
>
> [AEOLIAN:] "You sincerely think that the Duo-Art reproduces from your records so accurately as to satisfy one so well qualified to judge critically as your own mother?"
>
> [HEADING] *Duo-Art Reproductions Practically Perfect*
>
> [GRAINGER:] "Yes, surely. And when I myself hear the records which I have played at my best and then edited and corrected until they are my fullest musical expressions, I think to myself—'Ah, on the days when I play like that I am very well pleased.'"

[AEOLIAN:] "This is a fine thing you are saying for the Duo-Art, Mr. Grainger—tell me, will you go on record with the statement that the Duo-Art actually reproduces your playing even in such subtle things as graduations of touch and tone quality?"

[GRAINGER:] "That is a very legal sounding query, if you understand what I mean," replied Grainger, smiling—"Yes, I think the Duo-Art simulates every phase of my work, rhythm, tone, and all the rest. With reference to rhythm particularly, I am amazed at the absolute accuracy with which the instrument reproduces the artist's most personal characteristics."

Aeolian's targeted questions, the language of the overstated heading, and the caution of Grainger's answers are notable in this exchange. Grainger begins by offering a compliment on Duo-Art rolls, suggesting—along the same lines as many advertisements—that the playback of the reproducing rolls enables him to be present "in spirit" with his mother, or whoever else might listen to his rolls. The question posed next by the Aeolian representative does not seem overly particular or forceful; it mainly addresses the accuracy of the Duo-Art reproducing piano in successfully reading from the perforations in the Duo-Art rolls. But Aeolian's italicized heading, which appears before Grainger's response is even provided, and which appears to be intended to summarize the response to come, takes a huge leap from Grainger's eventual answer, advancing the claim that the Duo-Art's reproductions are "practically perfect"—something the pianist had never said at any point. In his own words, Grainger explains only that once the piano rolls have been edited and corrected, they are his "fullest musical expressions." For Grainger to state that he has edited his rolls to create an ideal musical expression is an entirely different matter altogether than stating that Duo-Art rolls actually reproduce the real details of his original performances at the recording piano.

These careful answers have not underscored the usual messages from reproducing piano advertising narratives: fidelity, touch, and the assurance that every single detail of the performance is faithfully captured and reproduced. And so, in this interview, Aeolian proceeds to press Grainger even more intensely on this issue than his earlier comments allowed: "Will you go on record with the statement that the Duo-Art actually reproduces your playing even in such subtle things as graduations of touch and tone quality?" In the transcript of this interview, Grainger's caution is palpable. The pianist's smile as he states that this sounds almost like a legal inquiry suggests that he is aware of what Aeolian is seeking from him and is considering how he ought to reply. His full response appears just as carefully worded as the loaded question from Aeolian. The interviewer asks Grainger whether the Duo-Art "reproduces [his] playing," including his touch and tone quality. Grainger provides the "yes" that Aeolian is looking for, but then qualifies his answer and chooses to use slightly different terms. He specifies that the Duo-Art "simulates" his work, and avoids using the term "reproduce," which Aeolian has used consistently.

Throughout this interview, even outside of this excerpt, Aeolian's questions have revolved around "reproducing" and "reproductions." Why would Grainger veer away from this term and choose to say instead that it "simulates"? In the pianist's final remark, after first stating that the Duo-Art gives a simulation of his work, he then attempts to offer the interviewer something more and provides a comment about how rhythm in particular demonstrates "the absolute accuracy with which the instrument reproduces the artist's most personal characteristics." Here, Grainger is willing to use Aeolian's "reproducing" terms to describe the Duo-Art after all, but only with regard to the accuracy of the rhythm. This is an important distinction because it suggests that—contrary to what the Ampico studio master quoted earlier suggested—recording artists did not always have the wool pulled over their eyes with regard to the real details of the capture system and its process for recording dynamics. Ultimately, pianists also earned money from these rolls, so their comments are not impartial, and it would not have served them well to directly criticize the company's processes and equipment. The goal of the reproducing piano was not an altruistic archival recording system; companies sought to make a profit, and the illusion of perfect fidelity sold instruments and rolls. Grainger's specific mention of the editing process contributing to the creation of his fullest musical expression makes clear the artistic reality of the collaborative studio work that went into a reproducing piano roll, much more than any scientific paper or patent could.

It was not only Percy Grainger who adhered to carefully chosen points when contending with Aeolian's leading questions. In an interview with Ernest Schelling, the American pianist offers comments similar to Grainger's about the fact that recording rolls for Duo-Art meant that he could reach people who may not have heard his performances:

> "Just what do you mean, Mr. Schelling, by reaching people through the Duo-Art? Do you feel that you are actually playing to them?"
>
> He was silent a moment. "Sincerely," he answered, "I think the Duo-Art reproduction of an artist's carefully prepared record will present

that artist at his best. For example, I consider that my interpretation of the 10th Rhapsodie which we heard a few minutes ago upon the Duo-Art was played as well as I would play it in one of my best moods. That makes my position clear, doesn't it?"

"You believe then," I queried, "that the Duo-Art reproductions retain the artist's personality?"

"Oh yes indeed. Particularly in rhythmical peculiarities, in tempi and in individuality of phrasing, the reproduction is startlingly perfect."[61]

In Aeolian's interview with Schelling, the interviewer asks pointed questions to obtain statements that align with their messaging, and Schelling is pointedly evasive, providing a carefully worded statement that avoids directly responding to Aeolian's question about whether Schelling is "actually playing to" people through his Duo-Art rolls. The pianist explains—phrasing his comments in a way that feels almost painfully cautious—that he feels Duo-Art rolls offer playing that is of the same quality as he would play when in his best mood, and that artists' edited and polished records present them at their best. Then, perhaps because he has not responded to Aeolian's original question, and because his position is in fact not very clear at all, Schelling adds, "That makes my position clear, doesn't it?" While Schelling's statement does in fact give us much more insight into his views, it suggests that his position may not necessarily align with his interviewer's. Much like Grainger, Schelling has stated very carefully that Duo-Art rolls produce quality piano playing, but he has not verified that they reproduce his own actual playing. Furthermore, his remarks comment on the fact that it is the careful preparation of the roll that presents the artist at his best, and not the recording process itself.

Both Grainger and Schelling focus on rhythm as one of the aspects they find to be most accurate on the reproducing piano—an affirmation that verifies the descriptions from technical staff, who confirmed that recorded rhythms were reproduced with little alteration. However, the two pianists can both also be seen beating around the bush when it comes to the overall fidelity of these reproductions. Aeolian's headings editorialize the pianists' responses, massaging their carefully worded responses into broad statements about "perfection" and "ideals." Ultimately, in spite of the ideas of living presence or scientific perfection of which companies worked to convince the public, reproducing piano technology relied just as much on a skilled human editor or interpreter as the pedal-powered player piano did. This human intervention was involved in a different stage of the creative process, and reproducing piano companies intentionally concealed the importance of these additional contributors, but Grainger's and Schelling's interviews suggest that pianists were not unaware of the reality of these compromises.

The story of the reproducing piano as an automation technology that could objectively deconstruct and reconstruct the playing of the great pianists was only that: a story. But untangling the threads of this story—the claims of "living presence" reconstituted by a machine, the invisible work of the editors slyly adding in dynamics while performers played through a trial run, the socially constructed concept of "fidelity" itself, the inventors who believed that artistry could be dissected and mastered with enough data, and the pianists who revised their piano rolls into musically satisfying playback instructions—reveals a story that is much more than a mere falsehood crafted to sell machinery. The story of automation technology as a tool meant to exert control over the building blocks of musicianship is a tale as old as time, one that stretches back to mechanical flute players and pinned music boxes and extends forward to sampled drum machines and holographic singers. But these reproducing machines are not bringing us increasingly closer to a fixed horizon of total automation. Much like in the case of the player piano, reproducing pianos did not erase the need for human involvement; they merely packaged human involvement differently. Pianists, technicians, and inventors did not turn the chronograph's velocity data into recreations of "living presence"; instead, they used automation technology to create something completely new and uniquely valuable: sophisticated musical programming that could be edited and shaped to convey a preprogrammed version of artists' "musical expressions" separately from a real-time performance event.

3. Synthesizers
Techno-Pop and Mechanization as Aesthetic

"We are not artists nor musicians. First of all we are workers."[1] This is a perplexing statement coming from Ralf Hütter, whose band, Kraftwerk, were honored with a Grammy Lifetime Achievement Award in 2014 for their pioneering work in the very art Hütter claims they did not practice. Only a few decades after advocates of the player piano had argued that the use of a new technology did not disqualify player pianists from being considered legitimate musicians, Hütter's statements, and the music of Kraftwerk, reexamined questions about the relationship between human artistry and automation technology. This, however, was precisely the band's aim: a rejection of the brand of artistic authenticity claimed by the transcendent virtuoso and the sweaty rock star, and the adoption of a new aesthetic that foregrounded artifice and technological integration. Kraftwerk were groundbreaking not only because of their pervasive use of synthesizers and sequencers, but because of their development of an aesthetic that intentionally blurred the boundary between humans and machines.

Situated halfway between the decline of the player piano in the 1930s and the rise of holographic performance media in the 2010s, the 1970s was a defining decade for the adoption of synthesizers, sequencers, and drum machines in popular music. Wendy Carlos's *Switched-On Bach* thrust the Moog synthesizer into the limelight when the album clinched the number one slot on the *Billboard* Classical Albums chart for three years straight, between 1969 and 1972.[2] While Carlos's studio album was a runaway success, musicians who brought the synthesizer into the broader landscape of live popular music performance were initially met with reactions ranging from fascination to contempt. In this story of synthesized sound, I examine two prominent electronic bands in the late 1970s that originated on opposite sides of the globe: Germany's Kraftwerk and Japan's Yellow Magic

Orchestra (YMO). Drawing on journalism covering these bands' early international performances and recordings, as well as concert footage and interviews with the musicians, I show how these bands led the way in a musical era in which the link between a musician's physical actions on stage and the sounds heard by the audience was weakened, or sometimes even severed. I also argue this disjuncture in no way meant that machines were replacing musicians, nor that these artists' musicianship was less human. The story of YMO's and Kraftwerk's popularity is one in which the relationship between human musicians and machines carved out space for new forms of creativity that have inspired generations of musicians.

With artists from British post-punk bands to Detroit techno producers to hip-hop pioneers looking to their work for inspiration, the legacies of Kraftwerk and YMO have extended far beyond techno-pop, as well as beyond their home countries, but these two bands have yet to be considered side by side, despite their similarities.[3] As two of the most influential techno-pop groups of the late 1970s, commonalities between YMO's and Kraftwerk's instrumentation, sound, and the role of their national identities help to underscore the ways in which the shifting status of synthesizers factored into the reception of both bands. Additionally, and perhaps most notably, both groups were formed in countries that had suffered defeat in World War II and had subsequently been occupied by Allied countries, which meant that during the decades leading up to these bands' creation, their respective home countries had been flooded with Anglo-American pop culture. Germany's and Japan's music scenes were dominated by American and British music, and most local musicians' work recreated the popular styles originating overseas. The international success of YMO and Kraftwerk went a long way toward legitimizing these musicians in the eyes of their home countries, particularly when their musical style departed from Anglo-American styles. Both Japan and Germany were beginning to develop a reputation as exporters of mechanical and technological products, and as such, techno-pop—with its similar aesthetics of mechanization and focus on high-tech automation—was able to underscore the connections between art and industry in the minds of international audiences.

Setting these two groups side by side, rather than focusing exclusively on the narrative of a single group, reveals different interpretations of synthesizer technology during its early years in popular music, helping to tease apart questions about whether these instruments were labor-saving machines, playful and expressive instruments, or a peek into a dehumanized future. In this chapter, I highlight the late 1970s as a flashpoint in the history of music technology in which automation became an aesthetic in its own right. This shift was part of a much larger upheaval in the 1970s, a symptom of a condition that futurist Alvin Toffler described as "future shock": a state of disorientation caused by an overwhelming rate of technological change.[4] The pressure to adapt to this new rate of change contributed to an increased interest in predicting what the future had in store, and this forward-thinking posture came to characterize much of the 1970s.[5] The themes that Kraftwerk and YMO explored in their music between 1974 and 1979—machines and humans, computers and video games, technologically driven cities and societies—added to broader cultural currents of curiosity and anxiety about the technological future. By foregrounding automation technology as part of their unique aesthetic, each band offered their own conclusions about where the relationship between art and technology was headed.

KRAFTWERK'S INTERNATIONAL SUCCESS

Kraftwerk's history begins with, and has been underpinned by, the efforts of its two founding members, Ralf Hütter and Florian Schneider, who met as music students in an improvisation course at the Robert Schumann Conservatory.[6] The two musicians created Organisation zur Verwirklichung gemeinsamer Musikkonzepte (Organization for the Realization of Common Music Concepts), an experimental music group with classical and electronic influences. They released just one album with RCA Victor, and the commercial failure of this record contributed to the dissolution of the group. Hütter and Schneider continued to make music together from 1970 to 1974 under the new name Kraftwerk, working briefly with a number of collaborators on three albums that tended toward free-form improvisation with experimental postproduction effects.

In 1974 Kraftwerk released their first internationally successful album, *Autobahn*. It was with this record that the band established a new aesthetic and a trajectory that would come to define their music. This aesthetic was decidedly modern, with a strong focus on themes of artifice and mechanization, from the futuristic appearance and timbre of the instruments the group used to the unemotional and satirical elements of their stage presence. The new Kraftwerk sound and image relied on heavy use of cutting-edge electronic technology; half-spoken, half-sung vocals; simple harmonies and lyrics that fit more in the realm of pop than experimentalism; and tidy, polished visuals on stage and in their album artwork. Although *Autobahn* was the group's fourth album, Kraftwerk sought to establish the record as a reimagined "year zero," serving as a first release linked to the group's new aesthetic.[7] The radio edit of the album's title track cracked the top five on both sides of the Atlantic, and Kraftwerk arranged a twenty-one-stop tour of North America. The musicians cut their long hair short and purchased sharp-looking suits before departing for the United States—a decision that quickly set Kraftwerk apart visually. Their image was distinct not only from the long-haired, grunge-inspired look of many rock groups at the time, but also from that of other experimental electronic groups in Germany, such as Tangerine Dream, who appeared in photos sporting jeans, hair past their shoulders, jewelry, casual shirts, and facial hair.

On tour, Kraftwerk fascinated audiences in the United States and Canada with their meticulous appearance and their pervasive use of synthesizers.[8] Although these instruments had been used previously in popular music, they were typically used more for occasional effects and not as the basis of an entire band. The synthesizer had spent the 1950s and 1960s as an esoteric tool used by academic musicians such as Milton Babbitt but had begun to be incorporated in popular music after Bob Moog equipped his synthesizers with a keyboard interface in the mid-1960s. Recalling their first American show in New York's Beacon Theater, Kraftwerk percussionist Wolfgang Flür described the impact of the contrast between the typical guitar bands who opened the concert and Kraftwerk's manicured appearance and futuristic synths: "When Ralf panned the thunderous sounds from left to right and back again over the whole stereo width of the stage during 'Autobahn', all I could see were open mouths and bewildered faces with wide-open eyes."[9] Kraftwerk's original twenty-one tour dates stretched longer and longer, turning into a three-month excursion. Headlines in Germany proudly described how the band was "conquering" and "electrifying" America, and took pride in the German band who had found success overseas.[10]

Descriptions of the group in the United States, however, revealed a different view of the band's sound: in contrast to the sound of rock groups—which were described as having a warmer and more emotional feel—Kraftwerk's futuristic equipment, tightly buttoned appearance, and strictly motoric beats came across as cold and unemotional, prompting frequent comparisons to puppets, robots, and stereotypically stoic and disciplined Germans. In 1978, the release of *The Man-Machine* prompted *Rolling Stone* to describe the album as "so antiseptic that germs would die there."[11] Many of these descriptors underscore facets of Kraftwerk's carefully sculpted aesthetic of automation and modernness, an approach that would be adopted by other bands in the late 1970s and early 1980s, by which point the synthesizer had become the defining symbol of musical modern identity.[12] Not only does Kraftwerk's early reception history parallel the discourse around the synthesizer as it became an emblem of modern electronic music making, but the group's aesthetic also played into the pop-culture interest in robots and other technologies that appeared in science fiction books and movies. The group's use of technology reveals an important contrast between the perception and use of synthesizers and those of player piano technology. Although the player piano and reproducing piano used mechanization and automation, respectively, to make music, the physical cabinet of these instruments, as well as the messaging used to sell them, worked to conceal the technological complexities of the instruments and foreground the artistry of the musicians who played them or recorded for them. Kraftwerk were working to do exactly the opposite, shaping their sound and image in a way that foregrounded the technological elements of their performances and minimized subjective human involvement. Automation had moved from a means of musical production to a foregrounded aesthetic choice.

AUTHENTICITY AND GERMAN AUTOMATA: KRAFTWERK MEETS LESTER BANGS

Both Kraftwerk's focus on technology and their connection to German cultural clichés proved to be important factors in the band's reception, and many interviews and reviews of the band center on one or both of these topics. The band had adopted their stiff, stoic image at the outset of their 1975 *Autobahn* tour in North America, and while the new aesthetic created a striking contrast with the appearance of long-haired rock groups and complemented the tidy appearance of their stage setup, it also fed directly into hackneyed American stereotypes of Germans.[13]

Kraftwerk's relationship with postwar national identity is a complex issue that extends beyond simply avoiding or playing into clichés. The ways in which the band gestured to stereotypes of stoic Germans in the 1970s were also ironic critiques, and viewing Kraftwerk's performances through the lens of this intentional and self-aware satire, both abroad and at home, is essential to understanding their aesthetic.[14] Kraftwerk was well aware of the dynamics influencing their reception, and journalists' tendency to label Kraftwerk as robotic scientists and efficient Germans frequently stemmed from comments made by the musicians themselves. In several interviews from the late 1970s, the band's founding members Hütter and Schneider played along with repetitive questions from journalists regarding their alignment with cultural clichés, but while these interviews occasionally reveal Kraftwerk's efforts to problematize simplistic national stereotypes, in other conversations with music journalists in the United States, the band seem only to exacerbate them. Ulrich Adelt has argued that Hütter's

German-language interviews tend to focus less on the complexities of German identity and offer more straightforward opinions than his interviews in English.[15] In conversations with English-speaking journalists, however, Hütter and Schneider can still be seen intentionally curating their group's German image and repeatedly sticking to particular talking points that helped to underscore the philosophy and aesthetic with which they sought to align themselves.

In 1975 the notorious gonzo journalist Lester Bangs interviewed Hütter and Schneider for a piece in the rock newspaper *New Musical Express*, and the article lays bare the tensions between Kraftwerk and the 1970s concept of rock authenticity. Bangs opened his article with freewheeling prose that linked subjects including the German invention of methamphetamine, the Third Reich, and the idea of a machine revolution. In his typically unapologetic and edgy style, Bangs asserted, "[Rock is] being taken over by the Germans and the machines.... The stupnagling success of Kraftwerk's 'Autobahn' is more than just the latest evidence in support of the case for Teutonic raillery, more than just a record, it is an indictment. An indictment of all those who would resist the bloodless iron will and order of the ineluctable dawn of the Machine Age."[16] Bangs invokes both the anxiety of a German takeover and a technological takeover for a readership consisting of American rock fans, arguing that *Autobahn* is not merely music, but a portent of a dehumanized musical future. While Bangs's provocative style of journalism is more flamboyantly forceful than that of his counterparts seventy years earlier, his rhetoric and choice of metaphors are reminiscent of comments on the player piano, which was also framed as an omen of the "advancement of this mechanical age ... the dawning of a new epoch, when machinery will ... be applied to uses hitherto deemed sacred from its invading banners."[17] The technologies themselves may have been different, but the anxieties and the language are striking in their similarity.

For many rock readers in 1975, who valued music that upheld a definition of authenticity grounded in gritty personal expression and visible instrumental virtuosity, Kraftwerk represented an opposing set of aesthetic values that destabilized physicality, presence, and emotional honesty. In the opening lines of his article, Bangs played directly into this conflict by painting an unsettling image of an imminent machine revolution to set up his interview with Kraftwerk:

> In the beginning there was feedback: the machines speaking on their own, answering their supposed masters with shrieks of misalliance. Gradually the humans learned to control the feedback, or thought they did, and the next step was the introduction of more highly refined

forms of distortion and artificial sound, in the form of the synthesizer, which the human beings sought also to control. In the music of Kraftwerk, and bands like them present and to come, we see at last the fitting culmination of this revolution, as the machines not merely overpower and play the human beings but absorb them, until the scientist and his technology, having developed a higher consciousness of its own, are one and the same.[18]

Bangs's opening intonation of "in the beginning" lends his narrative the gravity of a Genesis-like account of chaos into order, but here, the eventual arrival of order and control are hollow and are not human but imposed by machines. Humanity's loss of control is subtle and gradual in Bangs's narrative: first, humans think they have successfully controlled feedback; later, they merely seek to control the synthesizer; and finally, the machines have revolted, overpowering and absorbing human beings.

Bangs's description of scientist and technology becoming one and the same, however ominous it was meant to sound to his readers, appears at first to be not altogether different from Hütter's and Schneider's own descriptions of their group's concept. "Kraftwerk—Die Mensch Maschine" first appeared on tour posters in 1975, and the band's 1978 album took this same slogan as its title. The idea of "the man-machine" is pervasive in Hütter's and Schneider's interviews, but neither musician intends the term to reflect an erasure of the "man" in their music. Hütter emphasized this distinction repeatedly in his interviews. Explaining the group's adoption of "the man-machine" as a descriptor, Hütter said, "We call ourselves 'the man machine', which means the machines are not subservient to us and we are not the sounds of the machine, but it's some kind of equal relationship, or you might even say friendship between man and machine, and not opposed."[19] David Pattie argues that although these remarks were used as fuel for negative press concerning the band's robotic image, read from a different angle, Hütter was really only describing the relationship between artist and instrument.[20] Just as playing a guitar does not dehumanize the guitarist, Kraftwerk's mastery of technologically advanced instruments did not represent an escape from reality, but a meaningful interaction with technologies that were becoming an element of everyday life.

Bangs's framing of dehumanization as a frightening trend was only one possible interpretation of these aesthetic choices. The philosopher José Ortega y Gasset considered the issue of dehumanization and objectivity in 1925, arguing that while the music of the Romantic era had focused on the expression of personal feelings, twentieth-century music required a certain impersonal distance to appreciate fully. "Instead of paying attention to the sentimental echo of the music in ourselves," he wrote, "this music is something external to ourselves: it is a distant object, perfectly localized outside of ourselves and in front of it we feel like pure contemplators. We enjoy the new music concentrating towards the exterior. It is the music that interests us, not its resonance within ourselves."[21] Ortega labeled this focus on objectivity over private feelings *dehumanization*. However, the term was not intended to suggest a loss of human agency in musical performance, but a form of contemplation separate from an individual's inner emotions and anxieties.

Ortega's division between the Romantic emphasis on emotional disclosure and the twentieth-century focus on the contemplation of external objects is echoed in Keir Keightley's categories of authenticity in rock music, which he similarly divides along Romantic and modernist lines.[22] In an exploration of rock culture's long-standing preoccupation with authenticity, Keightley shows how rock's central definition of authentic art as direct, uncorrupted by commerce, and centered on sincere expressions of emotion has been wielded to draw dividing lines between rock and other forms of music that its adherents deem inauthentic. Authenticity itself is an extremely slippery concept with no fixed definition; however, critics, fans, and scholars have yet to tire of debating its meaning, owing in part to the important ways in which its ever-shifting definitions can shed light on the relationships between musicians, music, and audiences in a given time and place. Keightley shows how Romantic and modernist notions of authenticity—despite their conceptual differences—both celebrate the artist as a representative of an authentic self in their own way.[23] For the Romantics, that self was on a journey of self-discovery and expression of inner thoughts and emotions, while the modernist imagined artists as individuals who were driven to experiment, innovate, and reinvent both themselves and their relationship to their materials. Accusations of inauthenticity grounded in Romantic ideals often include the dismissal of "machine-made" music as fraudulent, or apply the term *dehumanized*—not in the way that Ortega defined it—but in order to imply that something of humanness itself had been stripped out of the very artistic practices through which we assert and express our humanity.

Lester Bangs's 1975 article delivers a scathing judgment of this perceived lack of humanity in Kraftwerk's electronic music before turning to the material from his interview with the band members. He opens with a rhetorical question that takes his doubts about "machine-made" music to their extreme end: "If any idiot can play [Chuck Berry's guitar lines], why not eliminate ... mistakes altogether, punch 'Johnny B. Goode' into a computer printout and let the machines do it in total passive acquiescence to the Cybernetic Inevitable?" This idea was not new in 1975, nor were the aesthetic judgments that accompanied it. In this article, however, Bangs extends his critique of machine-made music to include some bands that did not use synthesizers: "People used to complain about groups like the Monkees and the Archies like voters complain about 'political machines,' and just recently a friend of mine recoiled in revulsion at his first exposure to Kiss, whom he termed 'everything that has left me disgusted with rock 'n' roll nowadays—they're automatons!'"

Automata—the very same mechanical reproductions that had provoked fascination and fear for centuries—reappear here in 1975 as a knee-jerk criticism of a rock band that, from one listener's perspective, failed to display sufficient authenticity. The Monkees and The Archies are understandable as targets of censure for listeners who adhered to a Romantic concept of musical authenticity, since both were fictional bands that were assembled specifically for television shows. But why would Kiss, with their dramatic and intensely physical stage antics, be compared with automata? It may be that Bangs's friend perceived Kiss's carefully choreographed, spectacle-driven performances as lacking the same emotional disclosure and unpolished sincerity that others would find missing from the music of Kraftwerk. Ultimately, we cannot speak for Bangs's friend, but the most significant observation about this remark comes from the journalist himself, who followed up on his friend's comment, not with an affirmation, but with a more profound insight: "What he failed to suss was that sometimes automatons deliver the very finest specimens of a mass-produced, disposable commodity like rock." For a critic who is in the midst of an article that excoriates Kraftwerk for their use of machines, Bangs has made an important suggestion: that authenticity, emotion, and self-revelation aren't always the entire point, and that large-scale performances, even in the rock world, are ultimately a commercial enterprise. Kraftwerk was not attempting to meet the expectations of the 1970s rock aesthetic and coming up short; they were rejecting the myth of Romantic authenticity altogether. Although Bangs's article drips with cynicism about the work of his German interviewees, this insight aligns him more closely with Kraftwerk's aesthetic philosophy than perhaps either party would care to admit.

When Bangs finally does turn to the interview he conducted with Kraftwerk, the conversation opens with a discussion of the band's synthesizers and their studio, and Bangs wastes no time cutting to the chase on the idea of a technological takeover. "They also referred to their studio as their 'laboratory,'" Bangs explains to his readers, "and I wondered aloud if they didn't encounter certain dangers in their experiments. What's to stop the machines, I asked, from eventually taking over, or at least putting them out of work? 'It's like a car,' explained Florian. 'You have the control, but it's your decision how much you want to control it. If you let the wheel go, the car will drive somewhere, maybe off the road."[24] Across the history of automation technology in music, many critics have framed conversations around technology and music as a conflict that could end in the extinction of human creativity, but Schneider's analogy here suggests a different interpretation, in which technology is a tool, not an opponent with a conflicting agenda.

By far the longest uninterrupted quote in Bangs's interview comes from Ralf Hütter, to whom Bangs gives ample room for an extended explanation of the German cultural landscape after the war. Hütter describes how German music and entertainment was replaced with American imports and names his generation (and Kraftwerk in particular) as the first to create space for a new cultural identity. Comparing Tangerine Dream with Kraftwerk, Hütter points out Tangerine Dream's English name, and claims that they adopted an Anglo-American identity onstage. Kraftwerk, Hütter states, emphatically rejected this concealment of their German background: "We cannot deny we are from Germany, because the German mentality, which is more advanced, will always be part of our behavior. We create out of the German language, the mother language, which is very mechanical, we use as the basic structure of our music. Also the machines, from the industries of Germany." Here, the provocative nature of Hütter's statements about the German mentality and language does not seem to stem, as Adelt has argued, from any lack of facility in English. In this interview, we see Hütter and Schneider making intentional statements that help to solidify Kraftwerk's mechanized aesthetic and to link it to their identity as a German group who think like Germans and use German technology—unlike Tangerine Dream, who are painted as imitators of American artistry.

While Bangs seems on first blush to be the aggressor in his interview, both sides are tactically engaged in a drawn-out posturing contest, seeking to manipulate the flow of the conversation while avoiding direct engagement with the other's barbs. At the conclusion of the interview, both parties have exhausted their agendas, and the conversation reaches a stalemate. Hütter and Schneider are the ones to wrap up the conversation, and they cease their smilingly tolerant pretenses in doing so. Bangs describes how "Florian abruptly stood up, opened the window to let the smoke out, then walked to the door and opened it, explaining with curious polite curtness that 'we had also an interview with Rolling Stone, but it was not so long as this one. Now it is time to retire. You must excuse us.'"[25] Although both

Lester Bangs and Kraftwerk intentionally employed German stereotypes to underscore their messages throughout this interview, they each did so with different goals in mind. While Kraftwerk's generalizations are just as sweeping as Bangs's, they lay the groundwork for an aesthetic niche that the band could occupy in the minds of American readers and listeners.

There is one important difference between the way in which instruments like the player piano were framed as tools and Kraftwerk's electronic instruments as tools: Hütter and Schneider claimed that the synthesizers contributed back to their users, becoming more like creative partners. In a radio interview, Hütter described this reciprocal relationship in greater detail, and his careful explanation is worth quoting at length:

> To us it was quite direct to speak of The Man Machine because that's what we really are. It's the connection and cooperation of men and machines, because sometimes we play our machines, and sometimes they play us. It's like a dialogue: sometimes we switch on certain automatic machines and . . . they play very nice music . . . and we listen. We spend a lot of time listening to our machines, and then we change the programs and reset them. So it's like an exchange of ideas between us two. That's what "The Man Machine" is about, and also certain aspects in society where people are mechanically reproduced, or bought and marketed, or robots: the original Russian word "robotnik" means "worker." That's really our identity, what we are.[26]

Automation technology permitted the group to figuratively and literally take a step back and listen to a pattern looping on a sequencer in order to consider new layering effects or new rhythmic possibilities. While the group's working relationship with these machines was a valuable element of their creative process, placing automated technology on an equal footing with human musical artists also placed their aesthetic in direct conflict with the prevailing values of rock music at the time.

This symbiotic relationship between performers and machines modeled by Kraftwerk drew a great deal of attention, but also raised misgivings about the increasingly technological nature of music. In an interview for *Electronics & Music Maker*, Hütter stated that Kraftwerk's performance aesthetic was more than a natural relationship between musician and instrument, and onstage, their comportment was meant to evoke the robotic and mechanical: "So many people move or even jump around on stage these days and it's important for our music that we do not do this—our rather static performance is also necessary for emphasising the 'robotic' aspect of our music."[27] In the late 1970s, many saw this robotic style of composition and performance as a move toward fully mechanical music making that eliminated the interaction and connection between performer and audience. Kraftwerk lacked a relatable and engaging frontman, and in most performances, the four performers barely glanced at the audience. For many music journalists, this style of performance embodied the ultimate end of the gradual erasure of the pop performer as the star of the show—a claim that would recur years later in discussions of holographic performances.[28] Recurring anxieties about dehumanization and the erasure of the performer largely stem from Kraftwerk's emphatic rejection of 1970s rock ideals and the refinement of their mechanized aesthetic. Their simple harmonic progressions and repetitive melodic fragments were a target for critics, who argued that their music did little to convey a sense of musical skill or to communicate emotions of any sort.

Music journalists in the 1970s frequently positioned Kraftwerk as the polar opposite of the pop and rock performers typical of the time, and Kraftwerk did nothing to dissuade them. Of their carefully curated reputation and tidy shirts and ties, Karl Bartos recalls, "[Ralf Hütter] wanted to make it clear that Kraftwerk were different from any other pop or rock group and he wanted this image of a string ensemble. I didn't like it that much, I thought I always looked like a banker."[29] Bartos wasn't the only techno-pop musician who would go on the record complaining about his group's stuffy outfits. A member of the British group Orchestral Manoeuvres in the Dark (OMD), which in 1982 also sported short haircuts and shirts and ties not unlike Kraftwerk's, felt that his group resembled "bank clerks" onstage.[30] While this may well have been the case, the aesthetic choices of these bands in the United Kingdom also show how the tables have turned: British bands were now imitating a German band, rather than the other way around. In his work on new wave music in the 1980s, Cateforis describes the performance aesthetic of Tubeway Army several years after Kraftwerk had adopted their robotic postures and impassive stage presence in concert, and the many similarities are striking:

> Caked in a sheen of ghostly white makeup (applied at the show's behest to hide his acne) and black mascara, Numan presented an emotionally detached, unsmiling visage that would set a striking precedent for his future stage shows. Taken as a whole, the players' stiff postures, black uniforms, and "concentrated" approach to their instruments made them appear as if they were technicians in a work cubicle rather than musical performers.... They undertook their musical roles in a dispassionate, robotic manner. The presumably separate realms of humans and machines had bled over into one another in Tubeway Army's presentation. In many ways the band had left behind rock's normal expressive

domain, and was closer in spirit to the performing automata of the eighteenth century or Frederick Winslow Taylor's early twentieth-century conception of "scientific management," which equated industrial human labor with mechanical efficiency.[31]

Kraftwerk, OMD, and Gary Numan were not automata, of course—they were merely evoking them through their aesthetic choices, which then influenced their audiences' interpretations of synthesizers. In other words, the use of synthesizers did not necessitate stiff posture, straight faces, ties, and short haircuts. Rather, these aesthetic decisions were part of the interpretive framing of a new instrument.

David Pattie has demonstrated how this studiously stiff approach to performance runs counter to the notion of musical performances as public demonstrations of musicians' investment.[32] Performers conventionally prove their investment by displaying the very things that Kraftwerk omits: dramatic physical gestures, interaction with the audience, and visible evidence of emotion. By these criteria, none of the case studies examined in this book qualify as performances, but Kraftwerk perhaps run afoul of this definition by the largest margin. These performers are not simply machines or holograms that can be excused for failing to demonstrate investment that they cannot feel; they are human performers exerting effort to suppress expressive performance gestures and appear machine-like.[33] However, this very effort and intention underscores the fact that Kraftwerk are not performing any less than a rock band that screams and jumps onstage; they are just performing differently.

THE ROBOTS: A RECEPTION HISTORY

In 1977 Kraftwerk began to take their exploration of robotic performance aesthetics to another level. The group invested in a set of four lookalike dummies made to resemble each band member and outfitted their mannequins—which were sometimes referred to by the band and the press as "robots"—with Kraftwerk's trademark red-and-black outfits. On May 19, 1978, Kraftwerk released *The Man-Machine*; just over a week later, the robot-mannequins were presented for the first time on the TV show *19.30*, and they were later featured in the music video for the new album's first track, "The Robots."[34] At a launch party for *The Man-Machine* held in Paris, Kraftwerk set up their new robots and required the invited journalists to pose their questions to the lookalikes, which were equipped with prerecorded answers. Kraftwerk were pleased with the results of their setup—"Robots are patient; they don't object to anything or protest," noted Wolfgang Flür—the journalists, less so.[35]

"The Robots" became one of Kraftwerk's best-known songs, and the lookalike robots went on to become fixtures in the group's photos and performances, placed in front-row seats at concerts and intercut with footage of the human musicians in a music video. In the early 1990s the mannequins were replaced by actual robots whose arms moved in synchronized choreography, and these machines took the spotlight for performances of "The Robots," standing in for their human counterparts at center stage. These figures, despite their new movement capabilities, appeared less human than the mannequins they replaced, however, and featured intentionally exposed electronic innards. The four automata retained the molded heads of the musicians but no longer wore their signature red shirts and black ties. Below their detailed faces, the robots had only a featureless grey torso and hands connected by metallic arms with visible rods and wiring. In place of the legs that had once sported sharp dress pants, a pole supported the severed torsos. For years these robots delivered predictably perfect performances on Kraftwerk's tours. Their slow, sweeping arm movements—ironically much more dramatic than those of the human performers—and the projected visuals on the screen behind them have remained fairly consistent over the years. Perhaps the most interesting change was also the most gradual: the rubbery faces on each robot, replaced and kept up to date over time, gradually aged alongside those of their human counterparts.

When "The Robots" is performed in concert using these preprogrammed figures, there is no human performance taking place on stage. And yet across decades of concert footage showing performances of this song, audiences continue to cheer throughout the performance, as though the robotic stand-ins were capable of hearing and responding. The awkwardness that sometimes caused audiences at reproducing piano performances to hesitate and consider whether they ought to applaud a mechanical performance is absent here, perhaps because the original artists themselves are still nearby and still able to receive the cheers and applause, even if it is only from backstage.

During performances of "The Robots" in the 1970s, before the robots were visually engaging enough to hold an audience's attention for several minutes, the lookalike mannequins were sometimes seated in the front row, where the camera could occasionally show them seemingly watching the concert, while Kraftwerk themselves delivered an extra-stiff performance, standing as still as possible, jerkily moving their hands and arms to play their instruments, and barely moving their lips when singing. But in Kraftwerk's artistic and technological vision, the ultimate goal was never to deliver human performances that looked or sounded artificially stiff, and as such, these early performances of "The Robots," as well as the original mannequins, did not fully realize the group's ideals. Nor were Kraftwerk working toward a musical aesthetic in which human musicians no longer had a creative role to play. On many tours, the band members have walked off the stage and left the performance of "The Robots" in the automated hands of four mechanical figures. And yet the lookalike robots have never been tasked with delivering an entire concert. Their role is only relevant—and only meaningful—during a song about robots.

Of course, it is important to note that the robots' arm waving is not actually producing any music, and the four figures serve only as moving bodies to which the audience can anchor their gaze. However, these human-faced machines are important because they serve as a visual representation of Kraftwerk's "man-machine" concept. These robots, when they were given featureless torsos and poles instead of legs, were also intentionally given detailed human faces. Visually, the robots are hybrids, a synthesis of expressive human facial features and smooth metallic structural components. The concept of "the man-machine," too, is that of a hybrid: a synthesis of technology and expressive communication. In 1978 the best Kraftwerk could do was to deliver a stilted performance in which they behaved like robots—humans taking on stereotypical mechanical mannerisms. With sequencers and mechanized robots, Kraftwerk were in fact capable of delivering a performance entirely devoid of any representation of humanity. However, they chose not to, opting instead for a performance with human faces and expressive mechanical bodies that showed the relationship between man and machine and also gestured to the relationship that Lester Bangs forecasted, in which the musicians and their technology become one. When Kraftwerk stepped off the stage during performances of this song, and the rubber-faced robots began their synchronized choreography, the performance may have been automated, but it still wore the face of human creativity.

In the 2010s and 2020s, Kraftwerk reprised their red-shirted robots at some but not all of their concerts. During their 3D tour in 2022, I attended a performance in which the band exited the stage for "The Robots" and left the audience to watch an animated video projected onto a screen that featured the classic 1970s-styled robots. "Is this it?" I wondered, as the video and the song concluded. But just a few moments later, the large screen slid down to reveal the sharply dressed robots in all their glory (see figure 12). The four automata slowly gestured along with reprised instrumentals from "The Robots" while the audience whooped their appreciation, granting the

FIGURE 12. Kraftwerk's robots, featured in a performance of "The Robots" during a concert on the 3-D tour in 2022. Photo by author.

human members of Kraftwerk several minutes of reprieve before they filed back onstage to conclude the concert.

If early criticism of Kraftwerk's music played up fears of automation technology, warning that "the robots are coming," then with this song the robots are here, but the automated bodies onstage haven't taken over human musicianship; they have been built to explore the nature of a new type of music that benefits from a symbiotic collaboration between machines and human musicians. "It's feedback," explained Hütter and Schneider in 1975. "You can jam with an automatic machine, sometimes just you and it alone in the studio."[36] Ultimately, "the man-machine" was another way of describing a concept that connects each story in this book: the way in which musicians can use technology to access new forms of music making and, rather than stifling or replacing human artistry, how this partnership can amplify creativity. Rather than concealing or downplaying this partnership, Kraftwerk made machine collaboration a central feature of their performances and the language they used to describe their artistry. By taking up synthesizers and sequencers during a time when their meaning in music was still uncertain and foregrounding them in a technological aesthetic, Kraftwerk positioned themselves as part of an unfolding dialogue in the 1970s about the relationship between humans and machines in modern life.

YELLOW MAGIC ORCHESTRA: TECHNOLOGY AND IDENTITY IN EASTERN TECHNO-POP

Kraftwerk were not the only electronic group turning heads with a technological image and synthesized sound in the late 1970s, and they were also not the only such band to receive recognition in their home country only after they had successfully toured internationally and had hit songs overseas. After releasing their first album in 1978, the Japanese electronic trio YMO crossed the Pacific rather than the Atlantic to reach American performance halls in 1979, four years after Kraftwerk's *Autobahn* tour. Despite significant differences in their musical style, YMO were and still are frequently described in relation to Kraftwerk. Although the Japanese band can also be categorized as a techno-pop group, and there are similarities between the two ensembles, each band navigated issues of cultural identity and technology in markedly different ways, and YMO's career offers a contrasting take on the German group's version of a modern machine aesthetic.

Beginning in 1976, Haruomi Hosono, Ryuichi Sakamoto, and Yukihiro Takahashi, three Japanese musicians with separate but well-established music careers, connected as collaborators on a handful of projects ranging from Hosono's electro-exotica album *Cochin Moon* to Sakamoto's electronic fusion album *The Thousand Knives of Ryuichi Sakamoto*. The trio recorded their first album as a band in 1978, calling themselves Yellow Magic Orchestra. YMO's self-titled first record was quickly followed by a second album in 1979 named *Solid State Survivor*, released the same year in which the band undertook their first international tour. After an initially tepid reception in Japan, the timely pairing of *Solid State Survivor*'s release and a wave of positive reviews from their tour overseas quickly shot them to stardom at home as well as abroad. YMO scored hits in both Japan and the West, and *Solid State Survivor* found belated success, becoming Japan's best-selling album of the year in 1980—the year after its release. This recognition secured their position as fixtures in the Japanese musical landscape in the early 1980s. Drawing on the modern appeal of their synthesizer-heavy, futuristic sound, YMO were invited to partner with electronics and tech retailers such as Seiko and Sony, with the latter using a YMO song in a prominent cassette tape advertisement—a connection that bolstered the group's high-tech image.[37]

YMO's early history as a band contains several parallels with Kraftwerk's initial challenges and reception. Much like Kraftwerk's unremarkable reputation early on in Germany, initial reviews of YMO's performances in Japan were lukewarm, and it wasn't until the group gained international popularity that they drew acclaim in their home country. Similar to Kraftwerk and their desire to distinguish themselves in a sea of Anglo-influenced popular music groups, YMO was seeking to develop a unique style and sound that was distinctively Japanese but would also have appeal both inside and outside of Japan. "We needed something really powerful, something really new in the music scene," Haruomi Hosono explained to *Rolling Stone*'s James Henke in 1980. "We needed something that would be a bridge to the next pop form and that could be really powerful anywhere—in Japan, in the United States, in England, in Europe. I also wanted something that would be original to come from Japan. All the other musicians are following and listening to the music of the West and trying to do what they are doing."[38] From their early years, YMO were driven by a desire to innovate and to create something uniquely Japanese against a backdrop of widespread Western musical imitation, much like their German counterparts.

YMO began to record and tour just a few years after Kraftwerk had gained an international reputation, and the newer group frequently fell into the long shadow cast by the German band, with reviewers linking YMO to Kraftwerk in order to attempt to describe their sound. While YMO acknowledged their respect and admiration for their European peers, and although they used similar technologies and instruments, their musical style was far more distinctive than the comparison suggested. Whereas Kraftwerk's style in the late 1970s was minimalistic, emotionally detached, and austere, YMO's sound in the same time period trended toward rich textures, colorful harmonic palettes, danceable songs with infectious melodies, and stylistic traces of funk, disco, and guitar rock.

Although YMO would go on to become one of the most popular and musically influential bands to emerge from the 1970s, initial reactions to the group and their electronic music were mixed, even during their successful international tours. Japanese stereotypes, technological anxiety, cultural differences, and even Kraftwerk's stoic reputation factored into YMO's reception. An article from 1980 describes how YMO "has ushered in the age of the computer programmer as rock star . . . in which the computer does everything but wiggle its hips."[39] The idea that "everything" except for some sensual stage antics was automated in YMO's performances is an exaggeration that borders on absurdity, but other critics in 1979 and 1980 used similar language to overstate the role of the band's synthesizers and sequencers and to understate the musicianship of the performers. *Washington Post* performing arts reviewer Harry Sumrall scorned a November 1979 performance: "They call their music 'Technopops,' but other words also come to mind. Transistorized Tchaikovsky. Diode Disco. Robot Rock. Their songs . . . were reduced to a dreary, mechanical mush. The musicians never added a human touch to their playing. They preferred to let their gadgets do their work for them and, at times, it wasn't clear whether the men were playing the machines or vice versa. The result was an incessant drone which blurred any musical thought or substance."[40] Of a performance a month later in Los Angeles, a reviewer for *Variety* similarly commented: "Japan's Yellow Magic Orchestra brings an entirely new level of electronic existentialism to the icy German techno-rock school. . . . Things got off to an extremely remote start that night, via a series of seamless and ultimately tedious instrumentals which bore little if any traces of recognizable melodic warmth. . . . Crowd reaction was generally muted throughout, although the large audience complement of Japanese, who traditionally save their applause for the end of a concert, may have contributed to that."[41]

Both critics express opinions similar to those of Kraftwerk's detractors, regarding the perceived lack of separation of the roles of man and machine and the apparent dreariness of the concert. However, while the accusation of having "never added a human touch" to their performance may have been somewhat understandable at Kraftwerk performances, YMO's concerts bear little meaningful stylistic resemblance to those of their German peers. Although both bands made heavy use of synthesizers, YMO's stage setup, performance comportment, and use of nonelectronic instruments highlight significant differences between the two groups. Whereas Kraftwerk arranged themselves in a semicircle or a line behind a tidy grouping of completely electronic instruments, which they played with minimal physical movement, YMO made no effort to eradicate physical displays of human musicianship from their concerts. The band chose to build their synth-heavy sound on top of a drum kit and electric bass foundation, which set up a fundamental contrast between their band's sound and style and that of Kraftwerk, who eschewed traditional rock instruments.

Behind YMO's drum kit—an acoustic kit supplemented with rows of digital trigger pads that was positioned front and center onstage—Takahashi frequently grinned and bobbed as he played, sticks flying and cymbals flashing, creating a stark contrast with Kraftwerk's percussionists, who used minimalistic drum pads with drum machine sounds and played with little bodily movement. The three core YMO musicians toured with additional musicians in order to bring their studio recordings to live performance settings, including synth pioneer Hideki Matsutake, who programmed a massive Moog modular synthesizer live onstage, his skill at the towering instrument showcased by its audience-facing stage setup instead of stripped down and hidden; guitarists Kazumi Watanabe and Kenji Omura, who contributed blisteringly virtuosic electric guitar solos; and vocalist and keyboardist Akiko Yano, whose hands flew between the stacks of synths that flanked her onstage as she jumped, swayed, and clapped to the music as she performed. In short, YMO's stage presence, across numerous video recordings from their 1979 and 1980 performances, bore no resemblance to Kraftwerk's by any meaningful standard.

Although their music was still puzzling to some listeners, YMO's popularity was unmistakable. The *Sarasota Journal* reported that 180,000 fans had tried to purchase tickets for a show in an arena that seated only 10,000, but still labeled the band "mechanical be-bop," and attributed their fame to "a generation of Japanese teen-agers weaned on the beepings of TV computer games."[42] But YMO proved their popularity with young people outside of Japan as well and received an invitation to appear on the famous American music-dance TV show *Soul Train* in November 1980 (see figure 13). Hosted by Don Cornelius between 1971 and 1993, *Soul Train* typically showcased African American R&B, soul, and funk artists. With this appearance, YMO became the first and only Japanese band to appear on the show. The group chose to perform a cover of the 1968 funk hit "Tighten Up," as well as the dance-pop track "Firecracker."[43]

Although the audience in the studio danced along energetically and applauded the Japanese band's performance, when Don Cornelius joined YMO onstage, the typically smooth-talking host seemed uncharacteristically at a loss for words. After an uncomfortable chuckle, a long silence, and then another confused laugh, Cornelius cleared his throat and finally said, "In case you folks out there in television land are wondering what's going on. . . . I haven't the slightest idea." Cornelius's awkward puzzlement, and lack of knowledge about YMO's music persisted throughout his short interview with the band. After each group member gave their name, the host, grinning, confessed, "Yeah . . . if somebody offered me a million dollars to tell you somebody's name up here, you know I couldn't collect it, right?" After acknowledging the language barrier between them, Cornelius spoke briefly with Takahashi.

"Which of the current sounds do you think Yellow Magic Orchestra is close to?"

"Ah, it's difficult . . ." the drummer trailed off.

FIGURE 13. Yellow Magic Orchestra performing on the American television program *Soul Train* on November 2, 1980. *Source:* "Yellow Magic Orchestra—Tighten Up [+ Interview] Soul Train 1980, *Dailymotion*, posted July 1, 2015, by Felipe Cortney, https://www.dailymotion.com/video/x2waopy.

"You kinda have your own sound . . . is that . . . more or less . . . ?" Cornelius prompted.

"Someone said, ah, like Kraftwerk . . . Do you know Kraftwerk?"

"Of course! Hey, this is Big Don here, brother!" Cornelius paused to permit the audience a laugh and a cheer. "Nah, I'm not familiar with the record, and I'm sorry . . . but we're awfully glad to have you with us."

The host's unusual awkwardness was obvious during the brief exchange, and it is not entirely clear whether Cornelius meant that he was not familiar with Kraftwerk or with YMO's record. But the fact that Takahashi chose to reference Kraftwerk as the closest to their own sound highlights the challenge YMO often faced when positioning themselves in the popular music scene. Kraftwerk's name was immediately recognizable in connection with synthesizers, but beyond using similar instruments, the two bands' styles could hardly even be categorized within the same genre. "Firecracker," which YMO performed for the *Soul Train* crowd in this broadcast, cracked the top 20 on the US R&B singles chart in 1980—a chart on which Kraftwerk's tracks would never have appeared—and their funk-inflected style earned YMO a spot on a TV show that would have never featured the German musicians. On *Soul Train*, YMO turned out an energetic, danceable performance and pleased their audience with a fresh take on a familiar hit. In the context of the show's usual lineup of performers, however, YMO were extremely unusual. The *Soul Train* crowd were typically dancing to songs by Donna Summer, Stevie Wonder, and Kool & the Gang. Amid this collection of soul superstars, this group of slightly shy Japanese musicians took the stage to deliver a performance that blended funk and soul music with new electronic instruments and a generous twist of Japanese humor.

After being introduced by Don Cornelius, YMO started up the opening vamp of "Tighten Up," which included a screeched, tongue-in-cheek self-introduction in an exaggerated generic Asian accent: "Hi everybody! We are YMO! From Tokyo, Japan! We don't sightsee, we dance! You understand?" YMO took their parody of Japanese stereotypes further still, planting Youichi Ito, the band's manager, in the audience. Surrounded by fashionable young people dancing along to the music, Ito sported a clichéd Japanese salaryman's business suit, thick-rimmed glasses, a bulky camera around his neck, and a handheld sign that said "WOW!" This type of self-aware humor, which played up and poked fun at Western expectations of Japaneseness, was present throughout YMO's career, but their audiences were not always quick to pick up on it. Describing their reception on their first forays abroad in 1979 and 1980, Hosono recalls the misconceptions the band contended with: "When [YMO] did interviews, we were constantly getting asked the same questions. It was always things like, 'Does your music have a Zen influence?' or 'Are you influenced by traditional Japanese instruments?' This was 1980, so everyone still had a lot of misconceptions about what kind of place Tokyo was. YMO went global at the same time as the Walkman, so there was this image of technological progress, but it would get framed in a negative way."[44] The parallel here between YMO and Kraftwerk is significant. These threads of cultural stereotyping and technological progress were two of the biggest factors in both groups' reception outside of their home countries, and both groups folded these associations into the development of their aesthetic.

For YMO, these themes of technology and cultural otherness are visually melded in the album cover for the US version of the band's self-titled album, which used entirely different art than the Japanese release. Overseas, the cover depicted a woman in a Japanese kimono holding a paper fan, evoking Japan as it was viewed by Westerners. A tangle of multicolored wires

FIGURE 14. Album art for the US release of Yellow Magic Orchestra's self-titled album. Courtesy Lou Beach.

twisted outward from the woman's scalp—an Orientalist's Medusa with electronic snakes for hair curling above eyes hidden by a pair of oversized sunglasses (see figure 14).

Like the US cover art, the tracks on this album drape innovative synthesizer technology in self-aware exotica (including their cover of Martin Denny's Orientalist 1959 number "Firecracker," as well as a pair of pentatonically inflected tracks with Chinese titles), and this pairing sold well overseas. Like Kraftwerk, however, many of these decisions to take up and parody North American stereotypes of East Asians were intentionally ironic and sought to toy with cultural clichés, essentially exoticizing exoticism itself.

ELECTRONIC EXPRESSIVITY: YMO'S REINTERPRETATION OF SYNTHESIZER PERFORMANCE

While YMO took on a technologically focused image as a part of their aesthetic and identity, their approach to synthesizer technology differed significantly from Kraftwerk's, and these differing values are reflected in the music they recorded and performed. Kraftwerk framed their synthesizers as coldly modern machines and associated their music with technological landscapes and expressionless robots, but in YMO's hands, the same futuristic instruments could be blended with traditional rock instruments to produce music that was danceable, joyful, and full of energy. Not only were their arrangements more harmonically dense, their rhythms were more syncopated, their sound was more exuberant, and the group covered a number of popular songs on their first two albums, offering new takes on hits by artists ranging from Martin Denny to The Beatles. YMO never took themselves too seriously. The lightheartedness they showed during their *Soul Train* appearance turned up everywhere, from the comedy skits interposed between the musical tracks on the Japanese version of their album *X∞Multiplies*, to the types of sounds they sampled on their albums. YMO's self-titled debut album explored unconventional new materials in pop and electronic music and incorporated sound effects from video games from the late 1970s, including the popular game *Space Invaders*. By including these sounds, YMO was employing an aesthetic strategy similar to that of Kraftwerk, who incorporated the auditory experiences of everyday modern life into their music, such as car horns and the sound of the Doppler effect as cars sped past on the Autobahn. The video game sound effects sampled by YMO were drawn from arcade games that would have been a new and notable element of urban soundscapes in the late 1970s, and as such, they achieved a similar effect in establishing this album as decidedly modern. By including the melodies that featured in the gaming experience, YMO also called into question typical assumptions about what constituted authentic musical materials and authentic musical expression.

YMO's debut album opens with a track called "Computer Game 'Theme from The Circus,'" which comprises just under two minutes of beeps and bloops, which are sometimes rhythmic, and sometimes less than, but they are the very same sounds a player would encounter in Exidy's 1977 arcade game, *Circus*. Across the entirety of the track, only three melodic fragments emerge from the chaotic texture of arcade sound effects. The first is the simple start-up melody that players would hear at the beginning of each game of *Circus*. The second is the "funeral march" theme from Chopin's Sonata no. 2, which played when a character died, not only in *Circus* but in other early video games such as the 1975 arcade game, *Gun Fight*. This motif would have been familiar even to listeners who had never played video games, owing to its long history as death music in early twentieth-century film.[45] Finally, YMO includes *Circus*'s "success" melody—the well-known "Ta-ra-ra Boom-de-ay" vaudeville tune from the late nineteenth century. "Computer Game: 'Theme from The Circus'" was a highly eccentric opener for a pop album. Although there is an underlying beat included in some sections of the track, which gives listeners something to which they can anchor their sense of rhythm, other lengthy segments contain nothing but arrhythmic sound effects, creating the sense that this could be nothing more than a recording of an unknown gamer playing a round of *Circus*. Whereas the opening moments of Kraftwerk's *Autobahn* welcome the listener with a car's door slam and an engine starting up for a cruise down a German highway, YMO chooses to begin with a game. As the first track on YMO's first album, "Computer Game" was unconventional and unserious, both in terms of its musical materials (or the relative absence of them) and its structure—but this playful approach to sound and music making was entirely par for the course for the Japanese synth band.

East Asian studies scholar Michael K. Bourdaghs has argued that one of YMO's most radical moves was to undermine the opposition between authenticity and inauthenticity in music, proposing, in the same vein as Hugh Kenner, that in late twentieth-century media culture, that which is fake may be more real than the real itself.[46] Bourdaghs contends that YMO did not conform to the expectation that musicians' bodies serve as a visible expression of an interior truth and reads the use of Chopin's funeral march theme in "Computer Game: 'Theme from The Circus'" as an announcement of the death of the version of musical authenticity in which sounds are visibly coupled with a performer's body. Brian Currid has articulated this separation in techno-pop as "transmut[ing] the performance-centered traditions of rock by refusing any direct connection between performers' bodies and the production of sound."[47] Bourdaghs's and Currid's readings, however, are generally far more applicable to performances by Kraftwerk than they are to YMO, and Bourdaghs slips into the same pattern of interpreting the Japanese band in the long shadow of their German synth-pop counterparts. While Keir Keightley's definitions of modernist authenticity could serve as a good conceptual container for these kinds of visually decoupled, automated sounds, extensive performance footage from YMO's 1979 and 1980 concerts call into question this idea that the band's sounds were decoupled from their gestures at all. From video captured at their first US performances at the Greek Theatre in Los Angeles to the footage shot at the famous four-night string of shows at Tokyo's Nippon Budokan in 1980, rather than hiding their hands behind consoles like Kraftwerk, YMO's guitar fretboards, drumsticks, and keyboards are all on full, active display, and subject to frequent camera close-ups to showcase the band's skills.

Bourdagh's analysis of YMO suggests, contrary to evidence in concert footage, that the band "restricted the expressive body movements of the musicians on stage to a jerky minimum" and describes the band's cover of the Beatles' "Day Tripper"—which they performed frequently during their early years on tour—in terms that seem to echo those of the anxiously dismissive 1970s rock critics. He writes, "In YMO's hands, the song is transformed from an expression of teenage romantic angst into a surrender to the power of the machine. Their version features harsh mechanical drones, electronic squawks and squeals, and an overtly robotic, inorganic beat. At key moments, the melody grinds to an alarming halt, as if the machine had lost its power source."[48] While the studio recording of this track does feature synthesized sounds (Are they harsher or squealier than the distorted sounds of a rock guitar solo, or is the latter merely more familiar?) in addition to electric guitar and acoustic drums, its beat is anything but robotic and inorganic.

If the Beatles' "Day Tripper" is a tidy, concise, hook-driven classic, YMO's version is an ambitious, rhythmically complex display of virtuosity that puts on full display the band's chops in both electronic and rock performance. Rather than adhering to a clean 4/4-time rhythm, YMO play with beat displacement, hemiola, and the addition of extra beats in key measures in order to take what could have been a mechanically static groove and instead playfully destabilize the downbeat. For example, YMO insert an additional beat into a single bar early on in the chorus, resulting in a feeling of extended space between the lyrics "she was a" and "day tripper" that seems to simulate the stomach-dropping feeling of physically tripping into the second half of the line. Taken together, the complex rhythms of this cover keep the audience guessing, delighting and surprising listeners with the way in which subtle twists on the original overturn any suggestion of inflexible, machine-like beats. The album version of "Day Tripper" appeared on the 1979 album *Solid State Survivor* and was a staple of live performances on their first international tour. In the Beatles' hands, the song proceeded at a relatively laid-back 135 bpm. YMO's album version kicks the tempo up to a more driving 155 bpm, but in live performances, the band often pushed this to 163 bpm, and in their 1979 show at The Bottom

Line in New York, the song unfolded at a blistering 173 bpm, in which the shifting time signatures contribute to a frenetically virtuosic performance.

While the fiery solos and complex beats in "Day Tripper" showed that YMO had versatility and strong rock roots, the band was at their best not in a technically complex Beatles cover, but in their originals, which showcase their high-spirited, melodically driven musicality. Hits such as "Rydeen," "Tong Poo" (listed as "Yellow Magic (Tong Poo)" on the US release), "Behind the Mask" (which would later be covered by artists including Michael Jackson and Eric Clapton), and "Technopolis" all highlight their melodic creativity and energetic performance style, especially in live concerts.

In performance after performance around the globe in 1979 and 1980, in their *Soul Train* appearance, and in their approach to covers and original recordings, while YMO does interpret the synthesizer as a futuristic instrument that underpins a technological aesthetic, for this band, technology is more playful than inhuman. YMO used synthesizers and samplers to dramatically different musical ends than Kraftwerk. However, the German band's reputation loomed so large that YMO's chronological proximity and the simple fact that they incorporated synthesizers into their sound—even though their band was still underpinned by rock instruments, as well as soul and funk influences—meant that they were often interpreted as another unemotional, machine-like group, even when their performances were anything but. YMO's live performances advance the idea that in this new world of synthesized music, authenticity did not require sweaty demonstrations of emotion, but it also did not require forced solemnity or humorlessness.

. . .

The issues of authenticity and technology were not unique to electronic music in the 1970s, and accusations of inauthentic and inhuman music making still rear their heads in more recent electronic music genres. In twenty-first-century electronic dance music (EDM) performances, laptop performers have contended with the anxiety that they appear to be doing nothing more onstage than checking their email.[49] Work by Mark Butler has detailed the way in which DJs and laptop performers will often perform expressive engagement by exaggerating the physical movements necessary to adjust controls on pieces of equipment, sometimes seeming to turn small knobs with their entire body, communicating with their facial expressions, or dancing along with the music.[50] The DJs and laptop performers in Butler's study show that, even for twenty-first-century electronic musicians, audiences still consider physical displays of skill and emotion to be markers of authenticity, and a performer using electronic tools is required to navigate the use of visible performative actions in order to reassure their audiences of the fact that they are, in fact, undertaking musical labor.

The importance of Kraftwerk's and YMO's popularity connects directly into the story of Vocaloid music later in this book. First of all, YMO's massive success in Japan helped to generate momentum for the rich desktop music (DTM) scene that would flourish in the 1980s. It was this culture of creative amateur music making in the DTM community that would lay the groundwork for Vocaloid culture more than two decades later. Second, the reduction of physical performativity from live performances by Kraftwerk was an important precursor to the preprogrammed Vocaloid concerts that puzzled Western journalists in the early 2010s, and descriptions of YMO in the press during the group's early years bear frequent and striking resemblances to descriptions of Vocaloid performances.

Orchestras accompanying a concerto on a reproducing piano, techno-pop band members playing along with a sequencer, and instrumentalists serving as a backing band for a preprogrammed Vocaloid hologram are all participating in similar activities using automation technology, spread across a century of music making. Despite the comments from interviewers and reporters about dehumanization, and even when a group like Kraftwerk played up the technological elements of their creative process, the fears and criticisms about a loss of human creativity in music only increased the group's popularity, and this initially unsettling influence has crossed the boundaries of genre and generation. By prompting questions through performances that toyed with competing definitions of musical authenticity, Kraftwerk, YMO, and many later techno-pop bands were holding up a mirror to the cultural practices thought to confer authenticity on musicians in other popular music genres—even if this mirror was, in its own way, a hazy one.

4. Drum Machines

Synthesis, Sampling, and the Cultural Construction of Authentic Sound

A controversial drum machine advertisement ran in *Billboard* magazine on July 4, 1981. The top half of the ad listed the title of Elton John's new single, "Nobody Wins," accompanied by the descriptor, "The hit record" (see figure 15). Below that, an image of the LM-1 drum machine was paired with the text, "The drummer on the hit record." The advertisement was clean and simple, and the majority of the ad was whitespace, which meant that the image of the LM-1 itself was quite small, and very little detail could be seen on the drum machine's console in the photo. The advertisement's proclamation appeared straightforward: Elton John had replaced the human drummer on the hit record—who happened to be none other than the celebrated percussionist Jeff Porcaro—with a black box that bristled with an array of nondescript buttons and sliders. Dehumanization, indeed.

Journalist Rick Mattingly secured Roger Linn, the inventor of the LM-1, for an interview shortly after this advertisement was printed. During their conversation for *Modern Drummer* magazine, he noted to Linn that "some people are bothered" by this ad, and by the fact that it referred to the LM-1 drum machine as the drummer on Elton John's song. Linn's reply is typically mischievous, managing in the span of a few sentences both to directly disagree with and mildly insult the intelligence of readers who found the ad unsettling. "I liked the ad myself. When my ad guy came up with it I said, 'That's something that people will really like.' It's tongue-in-cheek. It's *not* the drummer on the hit record. Jeff Porcaro is the drummer on the hit record and it says so in the credits. Jeff thought of all the good parts and made the machine play them. All the LM-1 was doing was being his drumset." But the advertisement in question did not say "the drumset on the hit record." Had it done so, the ad would have simply presented the LM-1 as a new musical instrument, a source of new sounds and inspiration, and a

Drum Machines / 109

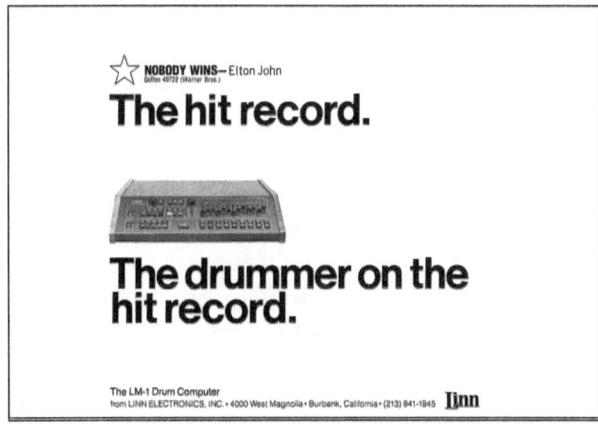

FIGURE 15. LM-1 drum machine advertisement that appeared in *Billboard* in 1981. Courtesy Eric Wrobbel Advertising.

new way to produce a great beat. Instead, it seemed to claim that the drum machine had substituted for a human musician.

Years later, Linn recalled the tension that the ad created, not just with magazine readers, but with Jeff Porcaro himself, and described how the *Modern Drummer* interview was meant to smooth things over. "I wasn't particularly sensitive at the time. . . . Jeff got pissed and wanted me to retract it. I did an article and said, 'I'm so sorry about that. It was tongue-in-cheek. Obviously, the LM-1 was not the drummer on the record. Jeff Porcaro was the drummer on the record.'"[1] Linn's tone looking back on this incident years later is certainly more conciliatory than it was in 1982, though the apology he imagines he gave in the original article appears nowhere in the pages of *Modern Drummer*. However, the advertisement and the controversy around it bring into exceptionally sharp focus one of the central questions about the drum machine that caused anxiety and uncertainty for amateur music lovers and professional performers alike: Is the drum machine best understood as a musical instrument or an automated drummer? Drum machines weaken the direct visual and physical link between percussion and percussionist, complicating the auditory relationship between a drum sound and its original source. If a drummer strikes a drumhead with a drumstick, the connection between the musician's body and the vibrating drum is direct and uncomplicated. If a drummer programs a rhythmic idea on a machine that has been engineered by one person and uses samples recorded by another person on a machine that automatically plays back the rhythmic idea at the push of a button, the clear relationship between musical sound and performing musician becomes troublingly complex.

The new music technologies examined in this book are among those that have blurred conceptual boundaries around performance and composition in music, challenging our ideas of what counts as an instrument, a concert, or a musician. Restrictive binaries can cause trouble in these conversations, because of the fact that they cannot properly encompass the nuances of a new musical tool that does not fit tidily into traditional boxes. In the situation involving the LM-1 on "Nobody Wins," the complicated categorization of the drum machine has a great deal in common with the reproducing piano. While the reproducing piano did occupy the physical place of a performer in the concerto performance conducted by Walter Damsroch, that did not make the reproducing piano a pianist. So it was with the drum machine on "Nobody Wins." As Linn noted—too late to spare him from offending Jeff Porcaro—the LM-1 was not credited as the drummer on the album, nor would it have made any sense for it to have been credited as such. The LM-1 was not a human musician with skills and experience. It was, however, a new type of instrument for creating rhythm tracks that many human musicians did use in the production of hit songs in the 1980s. Its popularity became so great that the machine became recognizable as part of the classic eighties pop sound.

During the 1980s, Michael Jackson, Queen, Stevie Wonder, Madonna, Prince, and many others wrote songs with preprogrammed drum tracks, and musicians who were in no way associated with the genre of techno-pop began to experiment with the new sounds and capabilities of drum machines that were entering the market. For some, these precise and polished beats were emblematic of the computer age, part of a fashionable modern aesthetic that aligned with Western society's embracing of new information technologies. For others, they threatened to spell the demise of human drummers, taking living, breathing musicians who had spent years mastering the drum kit and replacing them with a soulless machine.

There is a significant historical problem with treating performances on the drum kit as the standard for authentic human percussion and as the style of drumming that requires protection from technological incursions and cost-saving cuts. The drum kit—the very instrument that people feared drum machines had the potential to supplant—was itself a technological invention born out of the desire to streamline percussion performance into a more economical one-person operation. The drum kit is also the product of a gradual accumulation of mechanized technological developments that make it possible for a drummer to do things that would be humanly impossible without these mechanical aids. In the pit of theaters in the nineteenth century, multiple percussionists were hired to perform separate parts on the bass drum, snare drum, and cymbals or assorted sound effects. Closer drum positioning, plus the development of new mechanized pedal systems that facilitated increasingly efficient use of the feet for playing the drums, enabled managers to hire a single seated drummer to replace at least two standing drummers, and out of this economically expedient technological amalgamation, the drum kit was born. Across the generations of musicians that followed, the drum kit took on a life of its own, and innovations in both technology and technique resulted in a flowering of new styles of drumming that undergirded new genres of popular music, from jazz drummers' sophisticated grooves to the superhuman sound of metal drummers' high-tech, double-kick-drum setups.

In this chapter, I consider competing views of the drum machine during the most contentious decade of its identity crisis, drawing on perspectives from drum machine inventors, professional drummers, and music journalists. Some of these people were early adopters, experimenting at the boundaries of this new technology's capabilities and eventually shifting their artistic careers to incorporate the drum machine. Others saw potential in the drum machine for songwriting or producing and made use of the new technology in certain areas of their creative life, but continued playing the drum kit as they always had. Still others rejected the drum machine altogether. Beginning from the opinions and experiences of this diverse cast of music lovers, I move toward a consideration of the place of the drum machine in musical culture, arguing that these instruments were not a threatening technology with the potential to wipe out human creativity in the world of drumming, but a new tool that has been used in many creative and practical ways by different types of musicians. This perspective on the story of drum machines parallels Rousseau's assessments of the musical automata of the eighteenth century discussed in the introduction to this book. Like the mechanics whom Rousseau noted were the ones setting the fingers of the flute-playing automaton in motion, it was not the drum machine that played the drums, but the curious and creative people who took up the new device and sought to make new kinds of music with it.

FROM THE RHYTHMICON TO THE LM-1:
THE DRUM MACHINE'S ARRIVAL IN POPULAR MUSIC

Among the earliest twentieth-century efforts to create a drum machine was the Rhythmicon, the brainchild of composer Henry Cowell and inventor

Leon Theremin. The instrument, which premiered in 1932, had sixteen keys, each of which played a fixed number of notes per second. By playing multiple keys simultaneously, a user could produce polyrhythms that exceeded human capabilities in their precision and mathematical complexity. Cowell composed a pair of pieces specifically for the instrument, but despite its inventors' enthusiasm, the Rhythmicon and its intricate beat patterns failed to capture the imagination of audiences and other musicians and was relegated to the status of a museum relic.

The Rhythmicon was still an important stop in the developmental history of drum machines. The device was designed to augment the capabilities of human musicians, enabling them to perform rhythms that they could notate but could not perform unassisted, and as Schedel has argued, it also anticipated important developments in the interactivity of later automated musical instruments.[2] In *New Musical Resources*, published two years before the Rhythmicon was completed, Henry Cowell had already detailed a process for deriving sophisticated rhythmic relationships based on the overtone series.[3] The composer had been including these types of rhythms in his compositions for years, but patterns like five-against-thirteen were beyond the capabilities of human musicians to perform accurately. The player piano and reproducing piano did in fact have the capacity to produce the same types of complex cross-rhythms that Cowell was notating, and composers like Conlon Nancarrow used these instruments to explore precise polyrhythmic relationships in his *Studies for Player Piano*, composed during the late twentieth century. But Cowell was less interested in punching holes in a paper roll that would correctly reproduce a single fixed rhythmic sequence than in a new type of rhythm instrument on which a performer could play on the spot any rhythmic combination called for in a score or dreamed up in the player's imagination.

Thirty years after the Rhythmicon's debut, conductor Nicolas Slonimsky described the issue that the instrument's inventors had attempted to solve: "No matter what notation we may decree, human players will still be human—that is inaccurate, physiologically limited, rhythmically crippled, and unwilling to reform."[4] Slonimsky's harsh summary of human deficiencies was relevant not only in relation to the invention of the Rhythmicon, however; human limitations would continue to factor into arguments both for and against future generations of drum machines even fifty years later, as interviews later in this chapter show.

The music world had passed over the difficult polyrhythms of the Rhythmicon, owing to its disagreeable timbre and lack of applications outside of intellectual forms of music making, but the Wurlitzer Side Man, first produced in 1959, attracted much greater attention, commercial success, and controversy by offering simple preprogrammed beats such as the cha cha, rhumba, and fox trot, alongside which an amateur home organist or a cocktail lounge guitarist could find their groove. The American Federation of Musicians (AFM)—the largest musicians' union in North America—was quickly critical of the machine, fearing that it would cost drummers live performance opportunities. Ultimately, the AFM settled on a set of loose federation-wide rules around the use of the Side Man, leaving the details of rates and regulations up to local unions' discretion, while stating that under union rules, the device could under no circumstances be used to displace a live performer. The AFM's concern was understandable. Even Wurlitzer's choice of names lent itself to the idea that the machine was meant to stand in for a live musician; in music, a *sideman* is a professional musician hired to support a band or solo artist in live performances, and using this human label to name the machine created an impression much like the term *piano-player* had done when it described a device that promised to play people's pianos. Both piano-player and sideman were names for human musicians long before they were applied to machines, and both suggest the possibility that the machines could substitute for a human musician. As a labor union, the AFM's primary mission was to protect and augment the paid working opportunities of its members, and the Side Man was therefore a potential threat to its mission.

Wurlitzer, however, took the AFM's decision as an opportunity for promotion, seizing the occasion to run a full-page advertisement in *International Musician* that cheekily congratulated the AFM on its "progressive ruling."[5] In the advertisement, Wurlitzer laid out its own concept of the Side Man as an instrument that was designed not to oust drummers, but to aid solo musicians in practice and performance. "Wurlitzer created the 'Side Man' for the solo performer, organist or pianist," the advertisement read in large, italicized font. Below, in smaller print, the ad explained, "The concept of the 'Side Man' is that of an instrument designed to augment the performance of a solo performer. It does not—in fact, it cannot—replace a live performer. . . . It permits an individual to offer a superior performance. As a practice instrument, it aids amateurs and professionals in attaining and maintaining proficiency."[6] Despite its breezy confidence, the logic in Wurlitzer's advertisement copy came up a little short. If a human sideman's job was also to support or augment the performance of another musician, and the Wurlitzer Side Man performed the same function, then the main reason the device could not replace a human musician was neither ontological nor artistic: it was strictly because the AFM had ruled that it was not permitted.

Unlike the Rhythmicon, the Side Man was not designed to be a superhuman device that could do things that a "physiologically limited" drummer could not. The Side Man did not perform intricate polyrhythms. On the contrary, the appeal of the Side Man—whether it was tucked next to an organ in the living room for evening amusement or hauled along to a club gig when the drummer suddenly called in sick—was its simplicity and reliability. The Side Man did not play more complex or creative beats than a human drummer, but it could accompany a basic fox trot night after night without variation or fatigue.

In the 1970s a growing number of new drum machines hit the market, including Maestro's Rhythm King MRK-2 (1970) and Roland's Transistor Rhythm series (including the TR-33 [1970], TR-55 [1972], and TR-77 [1973]), which all—like the Side Man—featured a set of preprogrammed rhythms from which the user could select. The popular CompuRhythm CR-78 followed just a few years later in 1978 and was differentiated by a new creative feature: while this machine also came with its own array of preset rhythms like the earlier TR-series machines, CR-78 users could also program their own new rhythms from scratch on the CR-78, a feature that became increasingly popular on drum machines from the late 1970s onward. This generation of drum machines did not frequently appear on popular records, but this fact had less to do with any perceived deficiency in the instruments and more to do with the fact that they simply were not designed with studio work in mind.[7] Many of the analog sounds on these earlier drum machines had little in common with the sounds of the acoustic drums. The unnatural decay of the cymbal sounds and the click and hiss of drum sounds that were synthesized with white noise and square waves could provide a beat, but they certainly were not fooling any listeners into thinking they were produced on a physical drum kit. These 1970s drum machines were often used by musicians in their private songwriting processes. Guitarist Shuggie Otis described his introduction to drum machines in this decade as a creative tool: "I picked one up and it helped me lay tracks, but at first when I got it, I'd go home and I could play with it, like, any time I wanted to. I had my own little private drummer there who wouldn't talk back to me and he'd keep perfect time, you know? So that's how I wrote a lot of those songs — just messing around, you know, just jamming with that little Rhythm King."[8]

The LM-1 arrived on the music scene in 1980. It was not a product launched by a large corporation, but a small-scale operation by guitarist and engineer Roger Linn and his business partner Alex Moffett, whose last initials comprised the machine's name. Although the LM-1 was not the first programmable drum machine on the market, it combined programmable capabilities with two important new features: it offered digital sounds that had been sampled from physical drums rather than synthesized, and it enabled the user to tune each drum sound independently. These features pushed the cost of the LM-1 significantly higher than other drum machines, with its initial price set at $5,000, but they were also defining characteristics of an instrument that would grant its users greater control over the rhythms they were creating.

In an interview conducted by Rick Mattingly in 1982, the LM-1's coinventor Roger Linn opened the conversation by describing his intentions in developing his drum machine. "I invented the machine as an aid to songwriting," he said. "Any drum machine is a tremendous aid to songwriting in the sense that if you have a great sounding drummer sitting there playing while you're writing, it helps you write."[9] Linn's statement here, particularly as a guitarist, corroborates the statements of musicians such as Shuggie Otis.

"Do you sell more of these to studios or musicians?" Mattingly inquired of Linn.

"Musicians," said Linn. "Studios want to see a cost justification. Studios are more in the midst of selling something. Musicians are in the midst of satisfying their own emotional needs. They want something that's going to get them off, and this does."

Confident, slightly irreverent, and with a dry sense of humor that glints between the lines of his interviews and statements across various publications, Linn emphatically maintained his position that the LM-1 was created to serve musicians, not to replace them. The inventor insisted that his creation did not threaten employment prospects for drummers and believed the drum machine's primary use would be in making more realistic mock-ups of song demos. "I think the LM-1 will only replace other drum machines in situations such as home studios, where someone wouldn't have a drummer anyway," he explained to Mattingly. "In most cases where it was used on records, it was programmed by a drummer. So it didn't replace the drummer. In fact, in some cases, it got the drummer the gig."[10] Mattingly had at no point in this interview actually asked Linn to comment on whether the LM-1 would replace drummers, but the question seemed to be the elephant in the room, addressed by Linn before his interviewer even brought up the matter. Linn's confidence in predicting the future uses of his machine is clear. But technological histories consistently reveal an important truth about inventors' intentions: they rarely align tidily with how others choose to use those inventions.

MODERN DRUMMER MAGAZINE: PROFESSIONALS DISCUSS THE DRUM MACHINE

In a story often framed by contemporary journalists with antagonistic binaries such as "Drummer vs. Drum Machine," the drum machine's side of the supposedly antagonistic story is a rather unfussy history of creativity and innovation, in which inventors developed increasingly flexible and sophisticated engineering solutions to musical challenges, and musicians used them to explore new forms of creativity. But what did drummers have to say about the state of drumming in the 1980s as drum machines entered the scene? As it turns out, a great deal.

With decades of back issues in a collection that begins in 1977 and a readership of more than one hundred thousand drummers and drum enthusiasts, the American periodical *Modern Drummer* is among the most widely read and longest-running drum magazines in the world. Its year of origin—beginning publication just a few years before the launch of some of the most influential drum machines of the 1980s—also makes it an ideal platform with which to examine the unfolding story of the programmable drum machine's advent and its reception in the world of professional drumming. Starting in the early 1980s, there is no shortage of coverage in *Modern Drummer* of the launch of new drum machines, and a wealth of analysis and discussions populates nearly every issue for the better part of the next two decades. The details of these conversations offer a fascinating look at the way in which drummers navigated this particular instance of technological anxiety through discussions within their own community. As journalists, drummers, union representatives, inventors, and marketing executives bump their competing views against each other within the magazine's glossy pages, they provide a dynamic and revealing glimpse of the conversation around drum machine technology as it picked up speed in its early years. By tracking this conversation from 1981 to 1985, I show how the central debates include not only the typical uncertainty of the "identity crisis" years around a new technology, but also a great deal of optimism and creativity.

The earliest mentions of drum machines in *Modern Drummer* are brief, innocuous, and give no indication of the controversy that the machines would spark in the years to come. *Modern Drummer*'s first encounter with drum machines came in an interview with rock drummer and songwriter Simon Phillips in 1981, in which he mentions that he has worked with preset drum machines such as the RB-801 both live and in the studio.[11] In doing so, Phillips unwittingly becomes the first person to mention drum machines to the magazine's readership. At this point in *Modern Drummer*'s history, the periodical's interviewers seem to have either been uninterested in following up on this potential thread of the conversation, or merely omitted the rest of the drum machine discussion in favor of using the space on the page for other topics, since the topic of drum machines is dropped during the remainder of Phillips's interview.

In another 1981 interview, jazz drummer and record producer Harvey Mason discussed the LM-1 with casual positivity: "It's like a computer, and they put it in a drum machine, and you can program it to play a song all the way through with fills and everything, and it's great. You just do the homework at home, push a button and it plays. Sometimes when I produce, I don't want to play and do two things at once, so I'll use it. It sounds just like a drummer and it's incredible."[12] Mason's remarks suggest a function for the LM-1 separate from songwriting or subbing in for a drummer, describing how the device could free him up to give more of his attention to a different type of creative work. His remarks do not suggest that this drum machine enables a user to push a button and have it play a great beat. The first step of the creative process is important in his description: "You . . . do the homework." For Mason, relying on the LM-1 when he was focusing on his role as producer also required him to invest the effort to program a drum track that satisfied him as a drummer.

After its first unremarkable interview mentions, the LM-1 made no appearance in *Modern Drummer* for more than half a year, but in the February/March 1982 issue, the Linn Electronics device and its contemporaries became the subject of extended scrutiny and debate. The "Editor's Overview" for the February/March issue notes that there is "controversy" around the LM-1, describing it as "[a] threat to some, an aid to others" and observing ominously, "it appears the age of automation has started to hit home."[13] The editor's overview, as well as the titles of various write-ups with names that include oppositions such as "Friend or Foe?" and "Drummer vs. Drum Machine," seem tailored to play up the tensions of a potential conflict between machines and artists. The issue devotes a full eight pages of material to the details of "both sides" of the debate.

The choice of the "both sides" phrase used to set up the conversation around drum machines, in addition to the antagonistic binaries in the headings, shows that the editors have—intentionally or not—slipped into a classic but problematic rhetorical rut into which music journalists have repeatedly slid when they discuss new music technologies: that of reducing the reception of a technology to an overly simplistic for-or-against approach that draws up battle lines and pits earlier forms of music making and artistic values against the new: drummer versus drum machine, authentic versus artificial, for or against. Fortunately for musicians, these histories are never a straightforward, two-sided conflict, but nuanced stories of experimentation and creativity. This was true not only for guitarists, songwriters, and producers, but for drummers themselves.

The first writer in this issue to explain the LM-1 to readers is journalist Robert Carr, who offers a three-page description of the LM-1's functions and features.[14] Carr is careful and thorough, but not afraid to make an occasional jab at the complexity he perceives in the device ("a flurry of buttons and dials reminiscent of a bad dream after beer and pizza with pepperoni at 4 in the morning," he says of the interface). In the closing of his summary, Carr notes that he tested out the machine for himself, finding it "easy to use on the first try," but noting that it "would require a couple of days to a week's worth of work" to master the unit's subtleties.[15] This assessment of the LM-1, with its confident assertion of first-try ease, along with a low estimate of how long it would take to master the instrument, provides a striking parallel with the breezy assessments in early player piano advertisements: "A person who . . . never touched the keys of a piano in his life, can, in a few minutes . . . play upon the piano, with technical accuracy, any piece of music written for it. . . . [A] little practice is required in order to obtain the best results."[16] These kinds of assessments often dramatically underestimate or understate the intricacies of new music technologies, and for readers, this could be reassuring or worrying: just a few days to master a machine that could produce realistic-sounding beats? For hopeful producers with no skills behind a drum kit, this would be an intriguing proposition. For professional drummers who had spent decades to master the drum kit's subtleties, this could have sounded unnerving indeed. Fortunately, the magazine did not leave its readers to guess about what the professionals would think of the LM-1.

Modern Drummer also did not disappoint with the quality of musicians it found for its forum, but the opinions of the professional drummers may have fallen short for those who were looking for a scathing indictment of the machines that they feared were coming for their jobs. The interviews in the forum provide a great deal in the way of nuanced opinions on the LM-1. TOTO cofounder Jeff Porcaro provided the forum's first perspective to interviewer Robyn Flans, with Porcaro's interview followed by one with renowned session drummer Jim Keltner. While it appears that *Modern Drummer* may have intended the two musicians to have provided something of a counterpoint, framing "both sides" of the controversy as hinted at in the editor's overview, when taken together, Porcaro's and Keltner's remarks stand out for their balanced and open-minded approach to the device. Neither musician paints the device as a "foe," sometimes in spite of their interviewer's inflammatory questions and prompts. Flans, a music journalist who conducted some of *Modern Drummer*'s biggest interviews across decades with the magazine, presses both drummers in the forum with noticeably pointed questions that seem tailored to extract pessimistic views about the machine.

Midway through Flans's interview with Porcaro, after chatting about the situations in which he has used the LM-1, Flans notes that drummers are wondering how the introduction of the machine is going to affect them. Porcaro's response begins with a description of his vision of a future in which he can arrive at the studio with an array of his own "electronically perfect" sounds, plug them into a Linn machine or another drum machine, and play like he always plays. "You won't see a bunch of big drums, but if they can get it to where it sounds totally real and I can get the same dynamics and everything, what's the problem? I actually played it. It's my idea and I played it."[17] For Porcaro, in other words, the musical results matter more than the process used to obtain them, and the use of digital sounds does not strip drummers of the credit they are owed for their musical ideas.

In the same interview, Porcaro considers the new possibilities of remote work for drummers, who could transmit their work over a phone line thanks to the LM-1. This topic is one that the drummer returned to a second time in greater detail in a later interview in 1983, in which he still remained positive about the LM-1 but saved his most effusive enthusiasm for the way the machine could free up professional drummers to finish their gigs more quickly and get back to what they love doing: "Just imagine getting up and doing three dates within a two-hour period of time from your home and having the rest of the day just to play. Everybody's whole purpose in doing this is so they can make money and eventually they can retire and enjoy music."[18] Porcaro's comments on this theme in both of these interviews are important because of the way in which they directly counter the idea that automation technology was threatening to strip artists of meaningful creative work. Not only would the LM-1 fail to replace a skilled drummer like Porcaro; it enabled him to more efficiently turn out the kind of work that paid the bills but provided less in the way of artistic fulfillment. Although his comments might complicate the rosy, idealized image many nonmusicians and amateurs may have had about making a living in music, Flans attempts to smooth things over with a reassuring summary before moving on: "Still, all in all, Porcaro enjoys what he does and makes the best of it when he doesn't." Sometimes, if technology permitted, "making the best of it" could mean having the opportunity to take just forty minutes to input and transmit your contribution using a drum machine.

If Robyn Flans gave Porcaro space to enthuse about the LM-1 without philosophical debate in 1983, she wasn't quite ready to do so in the 1982 forum. In the earlier interview, after Porcaro explained his perspective on "phoning in" drum parts, Flans's response was to push back, injecting into the conversation an element of technological anxiety about the loss of human feeling. "Some might argue that it's a non-emotional approach to an instrument," she presses, following his story about remote work.[19]

"Yeah, I've heard that argument," Porcaro responds. "That was my argument, too."

Flans pushes the issue, asking him to defend his apparent change in position. "How do you justify that?" she asks.

"The song will justify that; the tune I'm doing will justify that. When a drummer programs it, it starts sounding like the drummer himself. It's his ideas that he picks to program, and then he can balance it the way he plays."

Rather than acknowledging his point about a drummer's sound and creativity coming through regardless of the type of percussion used, Flans continues to press for a critical take. "Could this be viewed as being anti-drummer in any way?"

"Sometimes, yeah," concedes Porcaro. "That's when musicians who are not drummers sit at home and use it.... [T]hen they call you to cut a track with real drums and they're not satisfied with how perfect your time is. It bugs drummers when somebody says, 'I want to use this instead of a real drummer.'" For Porcaro, then, the threat or foe in this situation is not the machine itself, but inexperienced users who have the idea that the machine could serve as a direct replacement for a skilled drummer. The irony of this conversation is that it is Flans who seems to be pushing this very idea. Porcaro's reasoning, however, centers on the idea that the ends justify the means, and that quality artistic work, whether it is on a drum kit or a drum machine, will speak for itself.

As the interview approaches its conclusion, Porcaro offers an emphatic reassurance that drummers do not need to feel anxious about the new technology, owing to the simple fact that the machine itself cannot drum. "Drummers shouldn't worry about it, though. In the musical realm, this machine cannot exist independently. A guy who plays drums is needed to program it. This is why we all learned our instruments.... This is just another piece of percussion—another instrument at our disposal. People shouldn't worry about it."

Here, Flans ceases to debate Porcaro and lets his reassurances stand. It is impossible to say for certain from the interview transcript alone whether her final question was meant to test the validity of Porcaro's opinions, or whether its noticeable brevity suggests it was asked with a sense of defeat: "Do you own one?" Flans asks simply, to which Porcaro gives his enthusiastic closing answer.

"Oh yeah. I'm even going to buy another one."

Jeff Porcaro rejected the "foe" side of the drum machine debate, but Robyn Flans had one more interviewee who could still spell out for readers the reasons the new technology should be feared and rejected. Jim Keltner eases into his discussion with Flans by narrating his thoughts about how he would "have fun with" the LM-1, using it to work out a concept and inspire new ideas to transfer to over to his work on the drum kit. "One thing about it is that it can get some grooves that no drummer can get," he says.[20]

This comment offers Flans a perfect opening to obtain her opposing take, and she is once again quick to pounce: "But if the machine is doing something that is beyond the capabilities of a human musician, isn't that a bastard art?"

The shock value of Flans's question covers up the substantial detour it forces in the conversation. What Keltner had just been describing was not the use of superhuman drum machine beats as a replacement for human drumming, but as a possible source of inspiration for his own drumming, which he would then execute separate from the LM-1. Flans's question abruptly turns the interview away from the generative possibilities of the drum machine as a practice tool and takes up the artistic status of the rhythms it produces.

Keltner's response seems a bit measured: "Well, you're talking about the computer age at that point. It's computer art, yeah."

Flans doesn't concede this point or acknowledge this switch from "bastard art" to "computer art," but swerves into a rhetorically charged statement that switches the terms of the conversation and introduces significant assumptions about the status of music created with drum machines: "Music needs more than technique—it needs feelings," she states.

"That you can argue real well," replies Keltner, perhaps catching that this indeed seems to have shifted from an interview to a debate. "No machine is going to have the soul of a human. I couldn't disagree with you in a million years on that, but you can't stop progress. Personally, I'm fascinated by the computer world, and at this particular time, I'm only looking forward to using the machine myself." Even under direct questioning, Keltner makes it clear that he does not land on the side of this argument that labels drum machines as a threat or foe, and he anticipates the potential of the machine as a source of new ideas for drummers as well as new forms of art.

The comments from the two drummers in *Modern Drummer*'s forum agree on a central point: that the LM-1, much like an actual drum kit, is a tool that serves the creativity of the person using it. Whether it was a source of inspiration for new beats, a device for storing and transmitting a drummer's work, or a new instrument in the creation of music in an emerging aesthetic style, the drum machine was not a harbinger of artistic destruction, but a device that held interest and appeal for drummers.

While Porcaro's and Keltner's comments did not offer the "both sides" contrast promised by the editor, the one point on which the two musicians did differ was on whether an experienced drummer was needed to program the machine. In Jeff Porcaro's comments quoted previously, he argued that someone who plays drums was necessary to program the machine, whereas Keltner's response when Flans asked, "Do you feel it has to be programmed by a drummer?" was simply, "Absolutely not. That's the beauty of it."

Flans did not share Keltner's optimistic enthusiasm: "But isn't that part of the fear of it?"

"Yeah, I could see where that could be a fear," Keltner conceded, but he immediately qualified his view of the nature of that fear: "The people who could be affected by that more than anybody, I would think, would be the guys who do jingles all the time. Like where there's a short little spot and it's always written out to a 'T.' ... The drummer could do it, but the arranger could do it as well because anybody can do it who can read music. If you want cymbal crashes, you simply overdub them."[21] Porcaro's and Keltner's opinions seem at first to contradict one another, with Porcaro arguing that a skilled percussionist is needed to program a drum machine, and Keltner saying that an arranger can get the same results by giving a written-out jingle to a drummer or by programming it directly into a drum machine. But taken together, these comments reflect a broader picture of drum machines' functions and use.

Porcaro's position highlights the fact that simply owning a drum machine will not produce high-quality or creative rhythm tracks on demand, and that someone without an ear for percussion would struggle to create a compelling groove, even on a $5,000 drum machine. In other words, for creative work that demands artistic skill, the gap between rhythms programmed by an experienced percussionist and an inexperienced layperson is huge, and a rhythm track programmed by someone without a trained ear for rhythm would never make the cut. However, on the opposite end of the spectrum, when it comes to work on which no creativity is required (or even desired), and a drummer is in the studio merely to hit drums in a pattern already dictated for a short jingle by a composer with enough musical skill to write out the part, then a drum machine could be used to help turn out efficient and precise work for a low-budget task that would never be assessed for its artistic merit.

This is the crux of most stories about a new automation technology in music: a new device—whether it is a player piano, a drum machine, or a piece of software for music creation—typically lowers the skill floor (the minimum level of skill required for basic competence) to create something simple and coherent, while also possessing its own separate skill ceiling (the highest potential one can reach through mastery). Player pianos permitted music lovers quicker access to a personalized play-through of a beloved Mozart sonata without years of piano lessons—lowering the skill floor—but the skill ceiling on the player piano was far higher than advertisers suggested or critics argued, and it was separate from skill at the piano keyboard, requiring careful study of pedal technique and air pressure management to deliver an excellent performance.

So it was with drum machines. With minimal experience and skill, a producer or a hobbyist could punch in a simple beat and use it in the studio or play alongside of it, but using these machines to their full artistic potential required practice, technical ability, and a high level of musicianship. An experienced drummer like Jeff Porcaro was able to learn the LM-1 and turn out quality rhythm tracks that showcased his own artistry and creativity. But in spite of the exciting possibilities inherent in the uncharted realms toward the skill ceilings of new technologies, the lowering of the skill floor is one of the main reasons behind a great deal of the automation anxiety that crops up in every generation. It has already been visible in criticisms of the player piano and synthesizer, and will rear its head again in later stories in this book. This shift leads some music lovers to fear that the mastery of hard-won musical skills will be overlooked. In other words, does quicker and easier access to simple drum work mean that the skills of the great rock and jazz drummers will be less respected and appreciated? This question encompassed a very real fear for some musicians in the 1980s.

BEYOND THE 1982 FORUM: THE DRUM MACHINE AS A CREATIVE TOOL IN THE EARLY 1980S

The 1982 *Modern Drummer* forum provided its readers with a glimpse of the perspectives of two professional drummers who saw artistic potential in the drum machine's future. Over the next three years, commentary on the drum machine flooded the pages of a magazine that had only recently even acknowledged the existence of the drum machine, and interviewers from this time period inquired about how drummer after drummer felt about the new technology. Taken together, the majority of drummers expressed views that were open-minded toward the new technology. Most professionals provided their interlocutors with an example or two of ways that they had used drum machines in recording or live performance, even if they weren't planning to make the new technology a cornerstone of their career. There were, of course, exceptions to the rule, and *Modern Drummer* did not shy away from printing some intensely negative predictions about the future of drum machines.

In February 1985, renowned jazz drummer Mel Lewis contributed an interview to *Modern Drummer* in which he discussed his long career and the current state of drumming, including the increasing use of drum machines such as the LinnDrum—the LM-1's successor—a topic on which the percussionist did not hold back:

> Now they have these machines, and I warned everybody about that at one time. If you could see some of the contracts up at the union right now, you would find out that half the jingles and half the record dates coming out are being done by one person with a synthesizer and a *LinnDrum* machine. As far as I'm concerned, the *LinnDrum* company should be blown up. I think it's a terrible mistake that rock drummers have been made to use these machines or have been talked into it, because they're being wiped out slowly. Some people will say that I'm nuts or that I'm full of crap, but they'll see. You can't wipe me out. A machine cannot do what I do.[22]

These are disconcerting words, coming from a veteran drummer whose band boasted one of the longest-running gigs in New York City.[23] Lewis painted a picture of a world in which hard-working musicians were being coerced into accepting a technological shift that was gradually destroying their craft and their livelihoods.

It might be easy to dismiss Lewis as out of touch: compared with other drummers, his comments seem over the top, and several decades later, rock drummers show no signs of having been "wiped out." However, Lewis's anxiety matters in this conversation because it underscores the reality that there will always be a range of responses to new music technologies, and that these responses reflect real experiences and feelings. In the identity crisis stage of a new technology's reception, there will always be early adopters, like many of the drummers interviewed by *Modern Drummer*, and there will always be those who have dedicated their careers to particular traditions of music making and have no interest in taking up and learning a new style or instrument. Mel Lewis passed away just five years after making these remarks, following a long and respected career as a jazz drummer. It should come as no surprise that this older drummer with a lengthy jazz career was not among the early adopters of drum machine technology.

In an important way, Lewis was completely right about the fact that a machine could not do what he did: a drum machine has no capacity whatsoever to improvise its own jazz solo, adapt with sensitivity and intelligence to its colleagues in performance, or maintain rhythmic fluidity within a groove. This is because it was never meant to do so, and the threat that Lewis was attempting to counter did not really stem from sentient beat-making machines, but from new aesthetic trends. A musician who could program a drum machine was indeed in competition with another musician who specialized on the acoustic drum kit, and for much of the 1980s, the fresh appeal and clean precision of the new drum machines' sound won over listeners in a significant share of the trend-obsessed world of popular music. But this is not the same as saying that musicians were replaced by machines. The history of popular music is filled with stories of the latest technologically driven sounds, instruments, and styles capturing musicians' imagination for a period of time and surging quickly to popularity, saturating the market until the next stylistic trend overtakes them. The drum machine opened up new musical possibilities and underpinned the rhythm tracks of entirely new genres, but for a drummer like Mel Lewis, the types of new possibilities the drum machine afforded were not only uninteresting but offensively antithetical to his artistic values. Mel Lewis may have believed himself to be fighting a battle against a machine takeover in the drumming world, but ultimately, the true battle lines were artistic: between a drummer nearing the end of a long career in a genre that valued collaboration and flexibility and other drummers who were exploring new aesthetic styles that centred on precision, independence, and new sonic possibilities. The most enthusiastic and innovative among the early adopters interviewed in *Modern Drummer* were percussionists more than twenty years younger than Lewis—musicians closer to the start of their careers, whose outlook still contained a streak of curiosity and optimism about new technologies.

Between 1982 and 1985, interviews with these early adopters added fresh approaches and insights to the unfolding conversation about the drum machine, with several musicians focusing on the machines' aid to precision in addition to their efficiency. One such musician was Terry Bozzio, best known for his work on numerous albums with Frank Zappa and Missing Persons. Interviewer Rick Mattingly inquired whether Bozzio used tracks from the Linn machine on the final mix of Missing Persons' *Rhyme & Reason*, and the drummer responded in a way that suggests that he was accustomed to assuaging doubts and worries about drum machines:

> I have no qualms about saying I used the *LinnDrum*. I think it's great, and I think any drummer who thinks it's stupid is being trivial and not very farsighted at all. This is a wonderful machine that was invented to help a drummer. In the old days . . . I would try to make sure I didn't breathe wrong or didn't adjust my seat in the middle of a hi-hat beat, because that could put a little dent in the time feel, and I would have to live with that dent every time I heard it on the radio. I know exactly what I want, so why not program it and have the machine do it? Of course, I've worked it out live. There's no problem with me playing it and executing it when we go on the road. But here's this great machine that saves hundreds of thousands of dollars in studio time by making it possible to get it perfect right then.[24]

Bozzio's comments blend a quick dismissal of the drum machine's detractors with a pair of arguments about accuracy and economics. The studio recording process enabled musicians to use multiple takes to get their parts sounding exactly the way they wanted them, but studio time was also expensive. If a musical idea originated with a performer and was entirely playable by that performer, then the drum machine was a tool providing artistic support. It enabled the musician to create exactly what they intended to create—especially for a recording, when listeners typically expected technical perfection to a much greater extent than in a live performance.

Narada Michael Walden took a similar approach to Bozzio, discussing the drum machine as a potential aid to his work. Walden, whose primary instrument is the drums, but who is known in the music world as a songwriter and producer as much as a drummer, did not consider it dehumanizing for a drummer to use a drum machine in order to achieve perfectly consistent timing when the trends and musical demands of a particular genre demanded it. "For top-40 music and commercial recordings, it has to be perfect. There can't be any rushes or dragging, and while I know a lot of folks are against it, a lot of it is done with a drum machine," he said. "I'm not saying I use it on everything, but I do when I know I want to be perfect."[25] For Walden, using the drum machine as an in-ear click track could support a musician's goals on the drum kit even when it was not audible on the final recording. He cited the work of producer Quincy Jones and drummer John Robinson—two award-winning musicians who contributed to renowned albums such as Michael Jackson's multiplatinum *Off the Wall* record—noting that Robinson was using a click track for all of the high-profile work he was doing at the time.²⁶ The drummer also noted in this interview that playing along with the unerring time of a drum machine required its own kind of skill: "It's an art to be able to have a click in your 'phones and make it sound natural," Walden explained, "and while a lot of people would say 'Man, that's a cop-out. How come you can't do it yourself?' Okay, you try and do it." Walden's apparent discomfort with broader perceptions of his work with the drum machine highlights the issue of anxiety around the use of automation technology as a tool. One reality that the introduction of drum machines revealed to many drummers, as Bozzio and Walden hinted, was that keeping perfect time indefinitely was nearly impossible, and that Slonimsky's descriptions of humans as "inaccurate, physiologically limited, rhythmically crippled, and unwilling to reform" was not entirely untrue.

Walden, and others who were interviewed on the subject in the 1980s, observed that drummers who struggled to keep time with the unrelenting precision of a drum machine often opted to give up or blame the inhumanness of the machine, rather than learning to keep better time. Many pieces of music do gain a great deal of their emotional power from intentional give and take in their tempo, whether it's the use of rubato in a nineteenth-century nocturne or the generous expressive fluctuations in a sentimental pop or rock ballad. Indeed, in both situations, a drum machine hammering out flawless timekeeping would not only seem out of place, it would create a feeling of emotionless rigidity. But for other musicians and other subgenres, rhythmic precision was not dehumanizing, it was highly desirable. Walden spoke of the pride he took in recording precision beats for tracks in genres like disco: "I wish more drummers would take commercial music more seriously and would take producing records more seriously," he said. "I'm telling you, it takes just as much know-how and discipline to play that music as it does to bash."²⁷

Disco, as well as later dance genres in which a perfectly steady tempo was essential—not just for the dance floor, but also for edits that relied on an unvarying pulse to splice together a new version of the track—benefited significantly from the precision afforded by drum machines. But detractors who blamed drum machines for forcing drummers into a constrained role in these genres, claiming they were micromanaged by click tracks, were missing some crucial historical context. In the 1970s, producers were already exerting increasingly greater control over the drums on their recordings. Producers had begun to use individual microphones for each drum in order to better isolate and adjust the sound, and eventually, drummers sometimes found themselves asked to record a beat one piece at a time (kick drum first, then snare, etc.) to grant producers full control over the final product. Beyond the studio, however, many DJs, editors, and hobbyists in the 1970s used turntable techniques and analog tape splicing to alter and extend their favorite recordings.[28] Drum machines were by no means the first time that percussion sounds were decoupled from their physical sources to afford greater control, and drum loop manipulation was used extensively in hip-hop and dance music cultures in the 1970s. Drum machines provided music lovers with the opportunity to construct a beat without spending hours of intensive work editing a tape with a razor blade and splicing block. In effect, then, the drum machine was a labor-saving device, but the labor saved wasn't necessarily that of the drummer, but of the fastidious producer who was creating the foundation for a new disco track or a creative editor who was extending a danceable drum break.

In the history of drum machines, the most memorable sounds and stylistic effects that musicians have created on these devices have often been precisely those that sound least like a person playing a drum kit. From the recognizable timbres of widely sampled machines like the Roland TR-808 and TR-909—which came to characterize the genres of hip-hop and techno respectively—to techniques like the stuttering hi-hat rolls that typified the subgenre of trap—an effect that was appealing in part because it would have previously required countless hours of laborious effort with a splicing block—listeners notice these sounds and techniques specifically because they sound so unlike acoustic drum playing, and that uniqueness is central to their appeal.

Mel Lewis and Jim Keltner were correct that many radio jingles and other tightly scripted recording work began to use the sounds of the LM-1 and other drum machines instead of acoustic drums in the 1980s. But it is an oversimplification bordering on an outright untruth to say that machines replaced human musicians. As the sounds of drum machines started to gain popularity, becoming part of the decade's sonic trends, numerous enterprising musicians—many of whom were drummers—picked up drum machines, learned to program them, and earned part of their income crafting electronic beats. Others chose to ignore drum machines and stick to what they knew best, and as musicians must so frequently do, shifted the makeup of their diverse portfolio careers to better align themselves with the available opportunities they preferred to pursue. Still others folded drum machines into their drumming practice, using them for creative inspiration, new timbres, remote work, rhythmic precision, polished recordings, layered textures, and an array of other possibilities. Over the course of years of dialogue and creative use, drum machines became much more than a mere stand-in for a drum kit on a demo, just as many other new music technologies' meanings and uses have shifted and expanded over the years.

"THE SONG WILL JUSTIFY THAT": CREATIVITY AND THE DRUM MACHINE, 1981–1985

In the response to Robyn Flans's forceful inquiries about the drum machine as "anti-drummer" and "bastard art," Jeff Porcaro argued that when a skilled musician used a drum machine, the work would speak for itself: "The song will justify that; the tune I'm doing will justify that," he told Flans.[29] Many drummers in the first half of the 1980s, during the drum machine's period of identity crisis in popular music, approached the new instrument with a great deal of curiosity and open-mindedness, often combining it with their existing tools and techniques. But how did recordings and concerts in which artists used a drum machine instead of or alongside an acoustic drum kit stack up? Three chart-topping songs from the early 1980s spotlight how drum machines were used to accomplish wildly different artistic and aesthetic goals and show how each artist worked creatively in ways that did not necessarily align with inventors' expectations for their instrument's use: The Human League's "Don't You Want Me" (1981), Prince's "When Doves Cry," (1984), and Phil Collins's "In the Air Tonight" (1981).

One of the first major hit songs to use the LM-1 immediately subverted Roger Linn's aims for his drum machine. The Human League released "Don't You Want Me" in 1981, a track propelled by an inflexible, robotic-sounding rhythm. Instead of drawing on the expressive features Linn had programmed into the LM-1, such as options for dynamics and more flexible-sounding "shuffled" rhythms, the band instead created an inflexible beat driven by relentlessly rigid hi-hat sixteenth-notes that extend across the entire track, including the times when a human drummer would have been forced to momentarily drop the hi-hat rhythm to execute a fill. The result, to Linn's chagrin, was anything but human, and the inventor expressed displeasure at the band's neglect of humanizing features he had worked so hard to develop.[30] But for The Human League—a synth-pop group heavily influenced by Kraftwerk—the inhuman aesthetic was entirely the point, and the LM-1 served their artistic ends well. Significantly, The Human League did not use a drum machine to replace their drummer; they simply never had one to begin with.

A longtime user of the LM-1, Prince never made the switch to newer drum machines such as the LinnDrum, not because of the cost of a new machine, but because the affordances of the LM-1 had become an important part of his artistic practice. The pioneering multi-instrumentalist and songwriter made extensive use of the LM-1 throughout the 1980s, crafting some of the decade's most memorable and unique percussion effects in popular music. Prince's aesthetic goals also undermined Roger Linn's script for the LM-1, but they differed from those of The Human League. Rather than creating a metronomically strict beat with no dynamic or timbral variance, Prince pushed the LM-1 to the opposite end of its spectrum of capabilities. He dramatically detuned the carefully sampled drum sounds and ran them through a range of filters that transformed sounds that were intended to be realistic into unrecognizable and highly individualized percussion elements. Prince's use of the LM-1 as part of this larger production setup shows an especially clear example of the way in which a creative musician can extend the affordances of a technological device beyond the boundaries its inventors imagined.[31] One of the most well-known examples of Prince's artistic alteration of these sampled drum sounds features prominently in "When Doves Cry," which opens with a brief solo flourish on the electric guitar, joined a few seconds later by the track's memorable beat. This beat features a hollow knocking sound that Prince created through extreme use of the LM-1's tuning controls. The distinctive timbre, and the effect it has on the track, is in no way intended to replicate a drum kit, but this difference is precisely what gives the opening of "When Doves Cry" its attention-grabbing quality.

The Human League and Prince are examples of skilled musicians who do not play drums as their primary instrument successfully using the drum machine to achieve particular artistic effects. Phil Collins's work with drum machines is of particular interest because he is also a professional drummer with a long career with bands such as Genesis and an outstanding singer-songwriter in his own right. "In the Air Tonight" was released in 1981 as the lead single from Collins's solo album *Face Value* and has its beginnings in a private songwriting session in the drummer's home studio. In an interview for *Mix* magazine—conducted by none other than Robyn Flans, more than twenty years after her contributions to the 1982 drum machine forum—Collins describes how he was offered a new drum machine from Roland while on tour with Genesis and initially turned it down, only to realize later that it could serve as a songwriting tool: "When I got back to find that I had a lot of time on my hands . . . I rang [Roland] up and said, 'Can I have my drum machine?' because I had to start writing some of this music that was inside of me."[32] Collins describes his process of setting up a simple, repetitive rhythm on the CR-78, finding some chords to loop, and pouring out the lyrics for "In the Air Tonight" on top of this basic foundation.

While the instrumental backing for the song's opening is hauntingly sparse and static, Collins has described the lyrics as containing "a lot of anger, a lot of despair, a lot of frustration."[33] But the gentle pops and hisses of the endlessly looping CR-78 do not clash with the intensity of the track's lyrical content; rather, the contrast between the two is central to the song's emotional power. Collins understood this. Though his drum skills are unquestionably up to the task of playing the simplistic two-bar loop in live performances, Collins has typically performed the song live either with a CR-78 onstage or with a sampled version of the original drum machine's sounds.

"In the Air Tonight" owes a large part of its enduring popularity to the iconic drum fill that shatters the song's tension as it approaches the four-minute mark. In live performances with a backing band, Collins choreographs the approach to the drum fill, slowly walking the stage empty-handed, singing into a headset microphone while the CR-78 loops, steering clear of his drum kit until just a few moments before the famous fill. A few beats before the moment arrives, Collins approaches the kit (often drawing cheers from the audience, who know exactly what is coming), takes a seat and takes up his sticks. His body transforms then, from patient stillness to dynamic power as he executes the iconic fill and performs the rest of the song on the drum kit himself. Robynn J. Stilwell has detailed a fascinating alternative seen at performances including LiveAid, where Collins performs the song solo at the piano.[34] Rather than building up to and playing the drum fill himself, Collins leaves the moment of the drum entry silent, turning to the crowd, who know exactly what to do, shouting out "Ba-dum ba-dum ba-dum ba-dum bum bum," a precise, audience-driven reconstruction of the recording. In this song, while Collins's fill is riveting and memorable, it is not merely the fill itself, but the contrast with the CR-78's feeble heartbeat, that gives the famous transition much of its strength. The aesthetics of a drum machine can be used to amplify an overall machine aesthetic in the hands of bands like Kraftwerk or The Human League, or to create new timbres when pushed to sonic extremes by artists like Prince, or they can be used as a songwriting aid. But in Collins's performances, their inhuman precision and distinctive timbre serve as an important sonic foundation that powerfully sets up the song's cathartic emotional and physical payoff.

CONCLUSION

In the decades that followed the 1980s, drummers continued to drum, and drum machines continued to be used by musicians in a growing array of different styles. In 2011 Roger Linn was recognized for his contributions to music technology, stepping onstage to accept a Technical Grammy Award at the Special Merit Awards Ceremony. He opened his remarks with a look back to the 1980s and a generous splash of his signature dry humor. "Sometimes I have a hard time explaining to people what I do for a living. I've found that the following usually works pretty well: I'll say, 'Do you remember back in the early 1980s when pop music started using drum machines and consequently lost all of its soul and humanness? Well,' I'd say, 'it's my fault,'" Linn said, pointing to himself, then pausing and letting his gaze fall to his notes with a wide grin splitting his face while the crowd laughed, thirty years after the fact, at the notion that one person's technological invention could have possibly dehumanized music. "I'm a very lucky guy," he continued with more solemnity. "To me there's nothing more personally gratifying than to create a new musical product, release it to the world, and watch how it affects this wonderful art of music. I can only make the brush; the artist chooses what to paint with it, and there have been some beautiful paintings that have been made."[35] A paintbrush is an artistic tool, not a substitute for an artist. Linn has made a number of such brushes in his career and continues to innovate, as well as to speak and teach at a range of electronics events.

Contrary to Mel Lewis's doomsday predictions, drummers were never "wiped out" by drum machines. The particular sounds of the LM-1 and other machines that gained widespread popularity in the world of 1980s pop music gradually became clichéd and fell out of fashion, replaced in turn by a steady procession of trendy new sounds and subgenres. Decades later, drummers have an unprecedented array of options available to them in their artistic work, from the latest electronic beatmakers to acoustic drums to the LM-1 and its contemporaries, which are now considered nostalgic machines from the past. In an interview for *Modern Drummer* with Death Cab for Cutie's drummer Jason McGerr, the longtime band member discussed his working process on the group's 2018 album *Thank You for Today*. McGerr explained that he viewed contemporary drumming as a layering of digital, analog, and acoustic sounds and described how some of his percussion textures include unconventional techniques like brushing fabric or layering handheld percussion, and others still use "older machines" including "a Linn LM-1."[36] The sounds of drum tracks in popular music are continually evolving, and artists are making use of a widening array of new "brushes," to borrow Roger Linn's term, to craft beats that continue to sound fresh and innovative. Drum machines were not a threatening technology with the potential to wipe out human creativity, but another new tool that was used in many creative and practical ways by different types of musicians and will continue to be used and developed in the music to come.

5. Singing Synthesis
Amateur Collaboration in Vocaloid's Creative Community

In the twenty-first century, synthesized speech is part of our everyday world. We listen impatiently as stilted voices slowly list far too many options in automated answering systems at our utility companies and banks. We chuckle when local street names are pronounced incorrectly in our GPS systems. And sometimes we are forced to speak slowly when we try to interact with virtual assistants who misunderstand our instructions. The experimentation and incremental advances that have led up to these experiences have taken place on a 250-year timeline, beginning in 1769, when Wolfgang von Kempelen hooked together bagpipe components and bellows to concoct a crude speaking machine.[1] The Austro-Hungarian inventor's device took twenty-two years to develop, and the completed version, finished in 1791, incorporated reeds, resonators, and swishing channels that helped to generate sounds from vowels to nasal consonants. A trained user could produce strings of complex words.[2] An untrained user could produce sounds that resembled an upset goat.[3] Kempelen's speaking machine was a product of the same Enlightenment interest in automata that produced Vaucanson's flute player and the defecating duck, when inventors worked to break down human actions into mechanical actions and construct devices to reenact them. Modern efforts in vocal synthesis, and the steadily improving systems they have produced, have accelerated over the past century, with significant milestones including Homer Dudley's Voder (short for voice operation demonstrator), first showcased in 1939, and Noriko Umeda's early computer-based, text-to-speech system, developed in 1968.[4]

Singing synthesis, naturally, posed a greater challenge than speech. IBM's John Kelly and Carol Lockbaum were the first to program a computer to warble a tune, using the IBM 7094 to sing "Daisy Bell" in 1961. Kelly and Lockbaum's achievement garnered recognition as a key moment in singing synthesis history, and "Daisy Bell" would later be sung by other computers as an homage to the IBM 7094, including the murderous supercomputer HAL in *2001: A Space Odyssey* and Apple's virtual assistant, Siri, when a user asked certain versions of the program to sing a song. The 1961 rendition of "Daisy Bell" was endearing, but the 7094 system came with a $3 million price tag in the 1960s, making the song more a technological stunt that was meant to showcase the computer's capabilities and less an attractive new avenue for music making.[5] After the IBM 7094, the timeline of advances in singing synthesis is striking for its brevity: less than fifty years passed between the time when Kelly and Lockbaum programmed their $3 million IBM system and the time in which a singing synthesis program called Vocaloid was available on home computers for less than $200.

Yamaha Corporation released the first version of Vocaloid in 2004. Developed in collaboration with researchers at Pompeu Fabra University in Spain, the program was intended to enable studio professionals to produce vocal tracks that could serve as background vocals or rough mock-ups of new arrangements.[6] With Vocaloid, users could input notes and lyrics into a software interface that resembled a horizontally scrolling piano roll, and the program synthesized these inputs into sung vocal lines. Composing songs in Vocaloid was a straightforward and accessible process, with simple click-and-drag functions for the placement and duration of notes. The program permitted the manipulation of pitch, rhythm, dynamics, timbre, breathing, and vibrato, enabling users to fine-tune the vocals and produce a more nuanced and expressive result. No notation skills were required, since the interface did not use a staff. Although the application was created for professional use, in the years following its release, amateur use of Vocaloid unexpectedly skyrocketed, and a dedicated community of creative users coalesced online. These amateurs collaborated on lyrics and musical arrangements but also on dance routines and animations for music videos, collectively producing highly polished musical and visual works, the popularity of which further snowballed the size of Vocaloid's user base and fan community.

Similar to using a plug-in for a digital audio workstation (DAW), users purchased third-party Vocaloid voice banks with distinct vocal timbres and a fictional character image on the box. While many voice banks and characters (referred to simply as "Vocaloids") have since been created for use with this application and are capable of singing in English, Japanese, Spanish, Korean, and Chinese, the most popular character by far since her introduction is a teenage girl named Hatsune Miku (see figure 16).

Miku's voice bank debuted in 2007 and took advantage of the significant improvements offered by the new Vocaloid 2 synthesis engine, released in

FIGURE 16.   Vocaloid character Hatsune Miku. *Source:* Art by KEI © Crypton Future Media, INC., www.piapro.net.

the same year. With her distinctively high-pitched and innocent-sounding voice and her long, teal pigtails, Miku quickly became the number-one fan favorite among Vocaloid users and listeners, selling forty thousand copies in just eleven months after the release (more than ten times the number of copies sold of the previously best-selling Vocaloid)—a spot she was still holding by her tenth anniversary, in spite of the introduction of many new Vocaloids whose voice banks boasted improved sound quality.

Singing synthesis's move from multi-million-dollar labs to home computers is more than a story about affordable software. Vocaloid's history is one in which a new technology's years of "identity crisis" offer a particularly clear example of the way that creative users can interpret and apply a new technological tool in ways far different than its creators intended. It is also a story through which we can trace the development of a new fan culture from its earliest days. Unlike many such cultures in which fans have admired a musician from afar, Vocaloid's story is one in which fans directly cocreated a fictional musician's music and persona through their collective creativity.

AMATEURS, COMPUTER MUSIC, AND ONLINE COMMUNITY

The success of professional groups such as YMO and Kraftwerk contributed to the growing popularity of electronic music in the 1980s, and this in turn led to a surge in amateur use of synthesizers and sequencers. It was in this decade that MIDI (Musical Instrument Digital Interface, a protocol that enabled communication between musical instruments and computers) was introduced, and this key development made DTM (desktop music) culture and computer music creation more accessible to amateurs. MIDI controllers facilitated a range of automated processes in computer-based music production, including sequencing and synchronization between instruments and software. Composers could input and edit musical data, which afforded them precise control over timing and pitch. MIDI also permitted users to program buttons and knobs in their computer's interface to automatically alter musical parameters over time. For example, MIDI automation could slowly shape the dynamics of a certain passage of music or gradually apply a frequency filter. Other automated tools simplified the process of composition in a digital environment, such as arpeggiators that automatically produced intricately patterned arpeggios, chord generators that enabled users without music theory knowledge to build harmonies with a single key press, and MIDI quantizers that adjusted inputted notes to align them with a given musical key or scale. These tools simplified or automated elements of the process of creating music, significantly lowering the skill floor for basic digital composition.

In the 1990s and 2000s, digital music making continued to grow and develop, and the introduction of hard-disk recording systems and advent of online music-sharing sites were two key developments that contributed to the growth of DTM culture. Across these two decades, recording and editing technology continued to become more affordable and accessible, enabling greater numbers of amateurs to acquire entry-level equipment. The commercial availability of affordable hard drives led to the development of hard-disk multitrack recorders such as the Akai DPS12 in the late 1990s. The DPS12 cost about $1,450—not a small purchase, but much more affordable for amateur music makers than previous technology.[7] In addition to its affordability, compared with earlier tape-based systems, hard-disk recording offered essentially unlimited track counts, no tape degradation, and nonlinear editing capabilities.

One of the most important shifts during this decade was the transition from storing musical data locally on computers to uploading and sharing creations on the internet as connection speeds increased and as new software and operating systems were released.[8] The launch of music- and video-sharing websites in the mid-2000s meant that amateur musicians had another place to encounter new work, upload their own, and connect with likeminded users without geographic limitations. The launch of MySpace in 2003 and YouTube in 2005 offered amateur musicians a new way to share and promote their music. The shift to online music sharing in DTM culture was highly significant: no longer were hobbyists limited to reproducing and distributing their work in physical form, a costly and laborious process in which creators' ability to share their music was limited by their contacts (including record shops that may or may not have been willing to put hobbyists' demo tapes on their shelves). Digital copies were infinitely reproducible, were rapidly transferable, and could be shared regardless of location.

Work by Keisuke Yamada has highlighted links between DTM culture in Japan and the Vocaloid community, tracing the story of Vocaloid developers' own roots in the DTM community and showing how their work referenced important technological precursors in musical synthesis.[9] In 1983, Yamaha introduced the DX synthesizer series, which included the extremely popular DX7. This instrument was the first commercially successful digital synthesizer in a market dominated by analog synthesizers, and it went on to sell more than two hundred thousand units. The DX series alleviated many of the functional restrictions that had previously posed difficulties for synthesizer users. They were compact, affordable, versatile, and—most importantly—had MIDI ports, allowing users to connect them directly to their computers. The DX7 was marketed to a wide range of musicians, including educators, home users, and professionals, and was taken up by both amateurs and prominent artists that spanned a wide range of genres, including Kraftwerk, Brian Eno, Enya, and U2.[10] As an homage to the enormous impact and legacy of the DX series, Hatsune Miku's design draws on color schemes and motifs from the DX7 and DX100 synthesizers. The panel on Miku's sleeve is a reference to the panel on the DX100, and the teal color of Miku's pigtails is the same hue as the teal on the DX7's buttons.[11] Miku's skirt also features a MIDI terminal design, suggesting that although Miku is designed to appear human, she is still the face of a synthesizer and an electronic tool that can be plugged into a digital system. These visual links underscore the relationship between earlier Yamaha instruments, DTM culture, and Vocaloid.

The story of Vocaloid's amateur community is a story situated at the ongoing intersection of three components: a Yamaha synthesis technology, an amateur music community, and a video-hosting website. On its own, Vocaloid is nothing more than another piece of music software intended for studio use. It was expected to sell only a few hundred copies. So then why was Hatsune Miku's voicebank able to sell tens of thousands of copies in its first year? The answer lies in the way this synthesis program was interpreted and used by a community of amateur musicians who took up Miku and Vocaloid synthesis technology during its early "identity crisis" years and created something unexpected. On the surface, the only significant new affordance that Vocaloid software offered was the ability to quickly synthesize sung lyrics with a roughly human sound. In the hands of tens of thousands of amateur users, however, "Vocaloid" became an entirely new genre of music that leaned into the capabilities and limitations of synthesized singing. Musicians who create with Vocaloid have built a fascinating body of work that comments on the relationship between humans and technology, highlighting the potential of creative coproduction using automated music technologies.

VIRTUAL SINGERS FIND A VIRTUAL HOME: NICONICO DOUGA AND AMATEUR USERS

Vocaloid's collaborative online history begins in 2006, when the popular Japanese video-sharing website NicoNico Douga was first launched, since it was this platform that would ultimately provide a space for the Vocaloid community to grow. My examination of this website focuses on its early years, from 2006 through 2013, which covers the rise of the Vocaloid fandom and also gives particular attention to the years before live video streaming was widely available on other platforms such as YouTube and Twitch.

NicoNico Douga resembled other video-sharing sites in its basic function—hosting videos uploaded by users—but with several key differences that were particularly well-suited to fostering the development of a user-focused creative culture. First of all, unlike YouTube, with its comment trees located in a section below the video, comments from NicoNico Douga's users appeared directly on top of the video itself, with each comment scrolling across the video at the exact moment in time during the video's playback at which the original poster made their comment. For example, if a user commented on an event occurring at the two-minute mark in a video, then as subsequent users viewed the video, the first viewer's comment appeared when they reached the two-minute mark during their own playback. This could result in what sometimes appeared to be a jumble of distracting text that partially obscured the video, particularly on popular uploads that received many comments, but the feature enabled a group of users separated by both temporal and spatial distance to react collectively in a format that simulated real-time interaction. On YouTube, live streaming only began to be rolled out to select users in 2011, the same year that streaming site Twitch launched, making NicoNico Douga's simulated simultaneity a uniquely attractive feature of the community in its early years.

In 2007, just a year after this video-sharing website launched, Hatsune Miku's voice bank was released for use with the Vocaloid 2 synthesis engine, and this timing made NicoNico Douga a convenient home for the nascent Vocaloid creative community. By 2013, approximately 30 percent of all uploads to NicoNico Douga were Vocaloid-related, a substantial share in the traffic of what was then one of Japan's top ten most-visited sites.[12] The rise of NicoNico Douga went hand in hand with the rise of Vocaloid, and the collaborative, creative bent of the NicoNico Douga community mirrored the development of similar traits in Vocaloid culture.

NicoNico Douga's simultaneous comment function facilitated a type of interaction that was unique to the Japanese website. In a music video, climactic moments in the song would often arrive accompanied by a wave of excited "oooohhhhhhhhh!" reactions pouring across the screen; moments of comedy could see users laughing together or collectively mimicking short phrases they found humorous. Sometimes, users would "sing" favorite lyrics together by typing along with them or engage with a song using symbols and abbreviations as a type of shorthand for emotional or thematic connections. Users also interacted with content on NicoNico Douga through the use of "tagging," a feature that enabled users to categorize videos. While uploaders could permanently assign up to five tags to their own videos, the remainder of the tags (up to a total of ten) were assigned and altered by viewers and could not be controlled by the original uploader. Originally intended for categorization, the feature was frequently repurposed by users for commentary purposes, including humor, criticism, and satire.

One of NicoNico Douga's defining characteristics has been its emphasis on sharing and collaboration in the production of creative content. Popular categories for video uploads that have persisted throughout NicoNico's history include "Tried Singing" (歌ってみた) and "Tried Dancing" (踊ってみた), and a significant amount of the website's content is user-created artistic content. Frequently, popular uploads to NicoNico have been shared and developed in ways far beyond the original creators' means. A group of musicians may arrange and perform a popular tune, while another user might choreograph a dance routine to accompany the new performance, and an artist may sketch a storyboard, which an animator using free animation software may turn into a music video. In the early twentieth century, player pianos were celebrated by musical amateurs as a democratizing technology that offered increased repeatability and customizable interpretive encounters with music outside of concert settings, but the way in which amateurs began to collaborate on NicoNico Douga went a large step further in democratizing musical creativity, owing to the fact that it facilitated not only private interpretations of existing works, but creativity on a collective and fundamentally generative level.

NicoNico Douga's open-source creative culture facilitated the collective production of Vocaloid works of a higher artistic quality than a single amateur artist could produce, and their diverse threads of influence and commentary have in turn contributed to a strong sense of community ownership, but Vocaloid fans weren't the first or only community to create and recreate layers of meaning. Lucy Bennett has documented similar trends in the area of digital fan studies, showing the way in which the boundaries between fans and producers can become blurred, with examples ranging from professional music to television shows having incorporated fan input into the final artistic product.[13] Bennett's ideas have their antecedents in the work of Alvin Toffler. The same author who in 1970 published the influential book *Future Shock* penned *The Third Wave* a decade later, a work that theorized the nature of the transition to a postindustrial society. Toffler advanced the idea of the "prosumer" (an amalgamation of "pro-ducer" and "con-sumer"), arguing that the lines between producer and consumer were becoming increasingly fuzzy as people gained access to new technological tools.[14] In many ways, Vocaloid music characterizes the prosumer dynamic to an even greater extent than professionally produced music and TV shows, because there is no single professional authority behind Hatsune Miku's songs, and fans who consume Vocaloid music and culture are frequently the same people who produce it. Rather than waiting for the release of official albums, Vocaloid fans cocreate new songs on

FIGURE 17. Scrolling user comments on a video. *Source:* "VOCALOID2 初音ミクに「Ievan Polkka」を歌わせてみた," *NicoNico,* posted September 3, 2007, by Otomania, accessed July 8, 2019, nicovideo.jp/watch/sm982882.

video-sharing websites, and apply new layers of meaning as they interact with them.

Owing to the extraordinary freedom users had to express themselves as prosumers on NicoNico Douga, memes and participatory responses developed over time and became fixtures in the Vocaloid community. These contributions functioned as more than temporary inside jokes, however; in this fan community, fans themselves had the agency to create meaning, and the meaning they created stuck. In September 2007, a simple, fan-made parody video of a roughly animated Hatsune Miku waving a green onion (*negi* in Japanese) to the beat of a Finnish folk song went viral (see figure 17), and became the first Vocaloid video to attain international popularity.[15]

Users began typing a series of green-colored letter Ys during certain parts of the song, to virtually join in a collective show of negi-waving participation. A comment placed early in the video's playback by one user provided helpful instructions for the uninitiated: "Y=negi." As more and more users joined in, the group collectively determined the moments in which the negi symbol should be used, and during these portions of the song, the screen would sometimes be flooded with strings of Ys. The practice caught on and spread, first to other videos on NicoNico Douga featuring the teal-haired Vocaloid, even though these videos had nothing to do with the negi image, and gradually to physical Hatsune Miku merchandise including art, keychains, and plush toys made by fans, irrevocably pairing Miku with the unlikely vegetable in the Vocaloid community's minds. Compared with the detached comment trees of YouTube in its pre-live-streaming days, the combination of expressive flexibility and time-specific linking in the Nico-Nico Douga comment system created a greater sense of communal viewership. The origins of Hatsune Miku's negi are just one example of how synchronized comment display simulated real-time collective interaction, which gave rise to unique forms of cultural expression that became an integral part of the developing Vocaloid community.

A growing body of scholarship focusing on users and their relationship to technological development has underscored the fact that users and technologies are always two sides of the same coin. Bicycles, telephones, video-sharing platforms, and virtual stars do not emerge from a vacuum, but are built to meet an existing need or create a new one. Often, however, the purpose for which a technology was originally designed is supplanted as users cultivate radically different purposes for it, particularly in the early stages of its development.[16] Musicians are central to this process in the evolution of music technologies. Trevor Pinch and Frank Trocco have detailed the evolution of Robert Moog's approach to modular synthesizer construction in the early days of the instrument's history, when several major instrument builders with differing philosophies were still in competition with one another.[17] Moog's developmental trajectory relied heavily on input from artists who were actively using the instruments, and musicians' involvement in this process significantly altered the finalized form of the synthesizer. Studies that focus on the blurred boundaries between the roles of producer and consumer, or designer and user, shed light on the ways in which new technologies in music do not emerge fully formed as a "threat" to musicianship, but are constructed in a gradually unfolding dialectical relationship. Similarly, in the world of Vocaloid—a creative community in which the roles of producers and consumers had a considerable amount of overlap—the meaning of figures like Hatsune Miku and of the artistic opportunities afforded by synthesis technology were co-constructed by members of the community.

The process of cocreation in the early years of the Vocaloid community is most visible in the earliest songs to gain significant popularity using Hatsune Miku's voice bank. After the simplistic green-onion humor of "Ievan Polkka" charmed online viewers with a cover of an existing song, a series of original songs uploaded to NicoNico Douga used Hatsune Miku's voice bank to explore what it could mean to bring to life the teal-haired girl depicted on the software's box. Three of these early songs went on to number among the all-time most-viewed Vocaloid videos, with more than one million views each, and all three of them were uploaded within the first month after Miku's voice bank was released, providing early input into the Vocaloid's identity and function during her initial period of "identity crisis." These songs were "恋するVOC@LOID" (*Koisuru VOC@LOID*, "VOC@LOID in Love"), by a producer called OSTER project; "みくみくにしてあげる♪" (*Miku Miku ni Shite Ageru*, "I'll Miku Miku You"), produced by ika and illustrated by KEI; and "Packaged" by producer kz with illustrations by redjuice.[18]

"Koisuru VOC@LOID," uploaded just two weeks after Hatsune Miku's voice bank was released, personifies Miku as a perceptive and eager musical companion, while still keeping the fact that she is a piece of software squarely in focus in the song's lyrics. The opening verse begins with lines that attribute feelings to Miku and create a sense of a relationship to the person who is using her software:

> That day I came to be by your side,
> Please, oh please, don't forget it
> When you look at me, you seem so happy
> So even though I'm a little shy, I'll sing a song for you.

A listener may be tempted to feel a bit of sympathy here for the shy girl singing for the user, but as the song continues, Miku seems to be commenting on the user's efforts to learn the new computer application, critiquing their skills, and then showing a little remorse for her harsh words:

> Don't mess with my parameters too much! Hey, you need to take care of me too!
> What about the attack, and things like that? You could pay a bit more attention to those.
> Don't cover it up with vibrato, those high notes are hard on me.
> I want to shine like I'm supposed to. Is this the best you can do?
> Sorry, I said too much. You're doing your best, I know.

Rather than producing a song that seeks to conceal the virtual nature of its vocalist, or hide the fact that the song was created using software with parameters like attack and vibrato, "Koisuru VOC@LOID" foregrounds the song's compositional process and uses the personified figure of Hatsune Miku to provide tongue-in-cheek commentary on the challenges of programming a quality vocal line.

"Miku Miku ni Shite Ageru," uploaded one week after "Koisuru VOC@LOID," is an energetic and intentionally lighthearted composition that rocketed quickly to fame as an introductory song to Vocaloid and Hatsune Miku. It became one of the six most popular Vocaloid songs of all time, with more than ten million views on NicoNico, and has inspired countless parodies, spinoffs, and derivative works. In "Miku Miku ni Shite Ageru," Hatsune Miku again sings in the first person, and the song's producer, ika, cheekily references her vegetable prop in "Ievan Polkka" in the song's lyrics as Miku introduces herself to a new user:

> I've come here from beyond the bounds of science.
> I didn't come with a leek but I wouldn't mind having one.
> Um, I wonder if you could hurry up and install me on your PC.
> Is something wrong? You've been staring at the package forever.
> I'll Miku-Miku you, and push myself to sing better.
> I'll Miku-Miku you, so you'd better get ready.

The reference to the leek was not lost on NicoNico Douga's viewers, who interacted with the video by appending floods of "Y"s through the comment overlay system, and the video's visual artist included a cameo of the same rough-drawn version of Miku that appeared in the video "Ievan Polkka" as a reference to the previous video's visuals. This early song contributed to a further sense of Miku's personification, including lines such as "The two of us are gonna make songs together," and "Please give me more to sing." The software itself provided no details about the pigtailed singer's personality; Hatsune Miku's image was nothing more than digital art of a girl on a box. These songs, however, attributed a sense of eagerness to Miku as a musical collaborator who could give voice to a user's musical ideas and feelings.

Compared with the frenetic enthusiasm of "Miku Miku ni Shite Ageru," in "Packaged," producer kz sets Miku's voice in a more laid-back arrangement, applying more reverb to the vocal line to create a spacious feel that reflects a sense of Vocaloid's broader potential. The opening lines ask, "The melody of the world, the sound of my voice / Does it reach you? Does it echo far?" The conclusion of the song provides an answer in the affirmative that suggests the potential of the nascent Vocaloid community to reach listeners with their music: "We can bring a smile to the world, you and me / It does reach out, it does echo far."

Although Vocaloid was first introduced as a substitute for a human singer, Hatsune Miku was not interpreted this way by the early users of this software, and these songs did not invite listeners to interpret Vocaloid music in the same way as a song performed by a human singer. These artists considered Miku's technological function alongside her fictional image and used her software to create music that explored a set of concepts: What does it mean for an automated voice to sing songs that we enter into this program? Can users interpret their relationship to this program as a partnership? How would an android singer understand the meaning of music she was asked to sing by human beings? In Vocaloid's early years, a significant share of the most-watched videos on NicoNico Douga featured songs that explore these questions, in which Hatsune Miku sings in the first person as a self-aware software program that tackled emotional, artistic, and existential questions from a uniquely technological perspective.

DERIVATIVE WORKS IN A CONTENT-SYMBIOTIC SOCIETY

By the early 2010s, the cooperative artistic work taking place in the Vocaloid community had become so influential that it attracted scholarly attention in Japan. A five-year music research project called OngaCREST (Core Research for Evolutional Science and Technology), undertaken at Japan's National Institute of Advanced Industrial Science and Technology (AIST), tracked and modeled the networks of collaboration through which Vocaloid works were being created and refined. The project produced applications that processed massive amounts of data from the NicoNico community and generated visualizations depicting the evolution of original Vocaloid works on an ongoing basis.[19] Of the project and the type of culture the OngaCREST team was seeking to promote, Masataka Goto explained: "We call a society in which relationships between humans and content and between past content and future content are rich and capable of sustained development a *"content-symbiotic society."* ... If a content-symbiotic society can be realized, media content can be richly and soundly created and used. ... Anyone will be able to actively encounter and appreciate content and, furthermore, enjoy creating content easily."[20] Goto's choice of terminology here reveals a shift in the meaning of user contributions in this content-symbiotic society. Rather than discussing the production of unique artistic works by individual creators, Goto focuses on the accumulation of vast amounts of content that exists as potential material for further creativity. In a model such as this, in which creating is a process of large-scale accumulation and development within a complex community, the OngaCREST team sought to develop tools for navigating this content that made it simpler and more efficient for creators to understand the relationship between previous works, and for listeners to discover new works—even those by lesser-known contributors.

To this end, in order to foster the continued growth of the content-symbiotic society that gave rise to the Vocaloid community, OngaCREST developed technologies that supported both content appreciation and content creation, including the music-browsing assistance service called Songrium.[21] Songrium was created with the Vocaloid community specifically in mind and has focused its analytical efforts primarily on music videos on NicoNico featuring Vocaloids. The application had analyzed more than 1.2 million videos as of 2025, the results of which have shed light on the processes behind the highly collaborative online culture, providing new insight into the role of derivative works and the small, individual contributions each video makes in the evolution of finished works. Songrium data showed that among the Vocaloid videos analyzed between 2012 and 2025, approximately 25 percent could be considered completely original songs, while approximately 75 percent were derivative works, such as instrumental covers performed by humans or animated dance music videos that interpret or develop the original in a new way.[22] This data generated by Songrium underscores the strong focus within the NicoNico Vocaloid community on collaborative artistic development through the contribution of derivative works. Additionally, by linking these derivative works back to their originals, Songrium works to highlight new contributions that otherwise might have been overlooked.

Songrium updates itself, collecting information from NicoNico, using algorithms to detect similarities between new and existing works, and linking them in an easy-to-use visual interface. Two-dimensional viewing modes such as Music Star Map assign each original song a position in 2D space, with arrows providing visual links to other songs that are musically similar, while its proximity to these other original songs is determined by how similar they are to one another. Songs with many musical features in common will be closer to one another in 2D space. Clicking on an original song brings up a detailed visualization of all derivative works associated with the original, in a format that resembles color-coded planets in orbit. The color of the circle representing each derivative work identifies the type of contribution as singing, playing, a music video, dancing, and so forth, while the size and speed of the circle as it revolves in its orbit indicate its popularity. The radius of each derivative work's orbit marks the date on which it was uploaded, with works created closer to the original orbiting closer to the center. For popular songs, these visualizations show a lengthy creative history: as of 2025, "Miku Miku ni Shite Ageru" had inspired no fewer than 1,394 derivative works (figure 18). As a data visualization project, Songrium offers not just a way to access Vocaloid music, but a way to understand the relationships behind its networks of cocreative meaning.

Masataka Goto's concept of a content-symbiotic society has key features in common with Henry Jenkins's definition of a "participatory culture," but while both terms highlight user-driven collaborative creation, Jenkins's definition places added emphasis on accessibility. Participatory cultures also feature low barriers to expression and engagement, as well as supports for creating and sharing new works. In this context, a website like NicoNico Douga cannot be thought of as a neutral platform. It consists not only of its code, but also of the social and cultural practices that shaped it and the

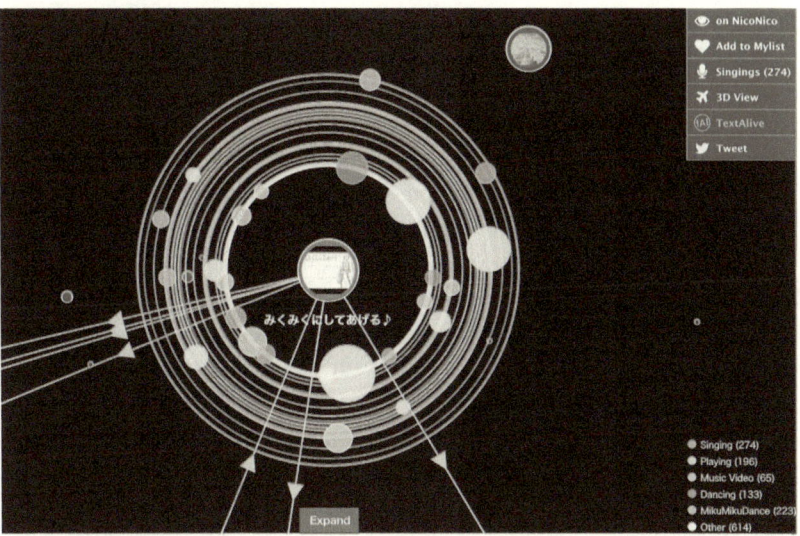

FIGURE 18. Music Star Map showing derivative works for "みくみくにしてあげる" (Miku Miku ni Shite Ageru). *Source:* AIST, "Songrium: A Planetarium of Music," accessed March 12, 2025, songrium.jp/map.

ways in which they facilitated (or inhibited) participation. A participatory culture is an environment in which members feel that their contributions make a difference in the culture and sense that they are socially connected with each other.[23] NicoNico Douga proved particularly helpful in facilitating this type of interconnected user engagement. The visibility of community-contributed comments and tags and the clear chains of collaboration involved in new works reinforced feelings of interconnectedness and cocreative meaning. However, while certain aspects of the website's interface made community interaction easier and more prominent than on other websites, NicoNico Douga was not designed specifically for the Vocaloid community's use, and the platform's features did not guarantee that a participatory culture would develop on it. New technologies can inspire certain uses, but these uses will only be adopted on a larger scale when they meet a need at a particular historical juncture. In other words, much like the use of Hatsune Miku's voice bank itself, the works produced on NicoNico Douga demonstrated that what creative users within a given subculture choose to do with the tools available to them matters more than the tools themselves.

Although participatory cultures existed well before the advent of the internet and websites like NicoNico Douga, Aaron Delwiche and Jennifer Jacobs Henderson have argued that a distinct new phase in participatory culture building, which they labeled "ubiquitous connections," began online around 2005—the same time period in which NicoNico Douga first launched.[24] Many researchers in the early 2010s approached this new phase of participatory culture with significant reservations about its potential for positive cultural impact, citing the challenges around intellectual property as well as regulations that stifled the potential of digital cultures.[25] Others contended that as new technologies continued to saturate every aspect of day-to-day life, the people who used them were becoming increasingly apathetic and isolated.[26] However, scholars studying fan culture argued that not only were fans in participatory cultures staying connected and engaged, they were also producing new narratives and works, maintaining and building social communities, and influencing professionally produced content through the additive impact of small contributions.[27] While this scholarship did not cover the dynamics of the Vocaloid community, the participatory culture that was emerging on NicoNico Douga was doing these very same things. Users were generating a large body of musical and visual work, growing their community through engagement on NicoNico Douga and fan forums, and collectively shaping content that ranged from music and videos to fan art that featured Hatsune Miku with her signature green onions. Even more remarkably, Vocaloid fans were undertaking their own efforts to remove barriers to creativity through the development of new technological tools made specifically for their community.

Animated music videos featuring dancing 3D models of Hatsune Miku began appearing on NicoNico Douga shortly after the release of her voice bank in 2007, and these popular videos racked up large numbers of views, but they were relatively rare, owing to the fact that they were more difficult for the average fan to produce than other types of content. Most Vocaloid community members had no experience with—or even access to—expensive and complicated 3D animation software. In response to this emerging need, Vocaloid community member Yu Higuchi developed an easy-to-use animation application called MikuMikuDance (MMD) and released it as freeware in February 2008. Compared with costly professional programs, MMD was simpler to work with, and its features were streamlined to meet the needs of users hoping to create their own Vocaloid music videos. In the years since its release, many other users have contributed to MMD by adding new Vocaloid character models, plug-ins, tutorial websites, and companion editing programs. The accessibility of MMD encouraged many fans to learn the software, and videos produced with the application became fixtures in the Vocaloid community—so much so that Songrium flags MMD submissions with their own unique color. As a creative tool, MMD had key similarities to earlier technologies like player pianos that could streamline aspects of a creative process that would have previously required extensive training and specialized skills. Much like player pianists who could pedal out a rough rendition of a favorite song without taking ten years of piano lessons, a MMD user could animate a Hatsune Miku music video of acceptable quality without needing to master the intricacies of a professional animation program. As a technological tool, MMD simplified many of the steps necessary to get a Vocaloid's 3D model moving and dancing on a fan's home computer, facilitating greater numbers of animated contributions to the Vocaloid community.

For fans, the authenticity of a Vocaloid song stems from the fact that it is not a manufactured hit produced by a team of professional songwriters looking to make a profit, but a creative work by a member of the community, for the enjoyment of the community. Fans feel a sense of agency and ownership in the bottom-up production process through which anyone can create and upload a song that expresses a feeling or an idea, help to develop the contributions of other fans, or bring an animated video to life. This rich creative ecosystem connects fans not only to the concept of Vocaloid, but also to a community in which amateur artistry has the supports it needs to flourish.

SUPERCELL: NICONICO DOUGA ARTISTS REACH THE MAINSTREAM

One of the most well-known success stories to emerge from the Vocaloid community originated on NicoNico Douga. Supercell has built a reputation as a popular music supergroup in Japan, with three studio albums, two of which have received Gold Disc awards from the Recording Industry Association of Japan.[28] In 2007, however, the eleven artists who would come together to form the creative supergroup were scattered across the NicoNico community, separately creating their own Vocaloid-themed works. The story of this group's origins and rise to mainstream popularity has been a favorite tale in the Vocaloid community and is an example of the potential within a content-symbiotic society for users to coalesce into collaborative groups capable of producing works that exceed the abilities of individual contributors. In December 2007, a musician who went by his given name of Ryo uploaded his first solo submission, "メルト" (*Meruto*, "Melt"). As a young songwriter without connections to instrumentalists or singers who could record his music with him, Ryo chose to use Hatsune Miku's voice over a synthesized drum and keyboard track, despite the fact that he had nothing to do with Vocaloid or its fan community prior to the composition of "Melt."[29] The song was a hit on NicoNico Douga, and as of 2025, the upload had accrued more than sixteen million views, situating it as the second most popular Vocaloid video of all time.[30]

The success of "Melt" would eventually lead to the formation of Supercell, but rather than the music attracting other talented contributors, it was actually the art that drew attention, and not for positive reasons. Ryo, lacking visual art to upload with his video, used an image of Hatsune Miku created by an illustrator called 119 without obtaining permission. When a NicoNico Douga user asked whether Ryo had approval from the artist to use the image, Ryo wrote to apologize and request permission. Not only did 119 consent, but the visual artist was impressed with Ryo's music, and the two began to collaborate. Gradually, Ryo and 119 added artists to their collective from 119's circle of friends and colleagues, and Supercell was born.[31]

Ryo continued to write and produce Vocaloid submissions the following year, continuing to collaborate with visual artists on three more videos, all of which quickly became popular hits on NicoNico Douga. In August of the same year, Supercell—by this time composed of eleven members—self-released an album containing twelve tracks, all of which featured Hatsune Miku's vocals. Shortly after the independent release, however, the group signed a deal with Sony Music, one of Japan's most prominent labels, and the album was remastered and rereleased in 2009, charting at no. 4 on the Oricon weekly albums ranking.[32] The timeline of Supercell's success story is remarkable: only two years had elapsed between Ryo's amateur upload of "Melt" and Sony Music's release of Supercell's self-titled album. From this point onward, Supercell continued to write music, but shifted away from using Vocaloid. In 2009, after the release of the group's first album, Supercell brought vocalist Nagi Yanagi—a NicoNico Douga contributor herself—on board to provide human vocals, replacing Hatsune Miku's synthesized singing in the majority of the band's songs until 2011. Since then, Supercell's albums have featured vocal performances from a handful of different human singers, rather than using synthesized vocals.

In 2012, however, Supercell released a track titled "ODDS&ENDS," which saw the return of Hatsune Miku as the featured vocalist for one song, much to the delight of Vocaloid fans. The lyrics that Ryo wrote for the Vocaloid singer made it a unique standout, which quickly secured an important place in the Vocaloid community. Whereas his previous songs had typically framed Hatsune Miku as a girl singing about themes such as youthful love, in "ODDS&ENDS" Miku speaks as a sentient computer program, in a style reminiscent of early Vocaloid hits. The song narrates her relationship with a user, whose creative ups and downs parallel Ryo's own artistic journey—a similarity that led fans to interpret the song as autobiographical. The lyrics of "ODDS&ENDS" are notable for the way in which they deal bluntly with criticisms faced by musicians who compose with Vocaloid, including negative comments on the sound of Hatsune Miku's synthesized voice ("My voice is a terrible offense to the ear" [なんて耳障り、ひどい声だって]) and accusations of musical inadequacy ("He's nothing but a fox borrowing a tiger's power" [虎の威を借る狐のくせに]). The song's lyrics follow the story of a creative person who is dejected and struggling to express himself, and in the first chorus, Miku offers him the use of her voice to communicate his ideas and feelings:

> So I'll lend you my voice! Some simply don't get it.
> My voice is a terrible offense to the ear, or so they say.
> But it'll surely become your strength! Let me sing and you'll see.
> Let me sing your very own words.
> Compose them and join them together, I'll cry out those words and feelings.

Eventually the musician finds success, gaining fame and a large following. He decides to set aside Vocaloid, stating, "Enough with this mechanical voice already! I'll be my own person!" (もう機械の声なんてたくさんだ僕は僕自身なんだよ) and comes to resent his association with Vocaloid. The song's conclusion, however, sees the musician undergo a change of heart about the value and impact of the music he made with Hatsune Miku's voice. He attempts to return to his collaborations with the Vocaloid singer, but to his dismay, he finds that he is struggling to make the program sing in the same way as before.

The official music video for "ODDS&ENDS" is set in a large, dimly lit room cluttered with out-of-date technology. CRT television sets, old computer components, and bulky phone handsets lie discarded among heaps of nuts and bolts—odds and ends, per the song's title. The opening moments consist of a series of close-up shots of various technological devices as they blink and spin, accompanied only by a gentle beat and a looping, rising melodic figure. The music video's main character is a tiny robot with a boxy, metallic body, who discovers a handheld screen displaying Hatsune Miku's face among the piles of devices (see figure 19). Enamored of the singing girl, the robot seeks to retrieve Miku's screen and prevent the display from running out of power, continuing to labor at this task as the narrative in the song's lyrics unfolds.

FIGURE 19.   Toy robot pulling a display showing Hatsune Miku's face in the music video for "ODDS&ENDS." *Source:* "【初音ミク】 ODDS & ENDS - PV Full Ver.," *NicoNico*, posted August 12, 2012, by まさゆき, https://www.nicovideo.jp/watch/sm18592204.

The viewer frequently sees the two-dimensional image of Miku lip-synch to some of the song's lyrics as the tiny robot pulls her screen around the room. This perspective frames the image on the tiny handheld screen as the singer in the music video, while the band's human instrumentalists perform in the background of the room, seemingly oblivious to the struggle unfolding amid the piles of odds and ends. However, for nearly the entire music video, the viewer only sees the bottom half of Miku's face, and often just her mouth. The fact that Miku is singing in the first person seems, on the one hand, to confer human agency on her, enabling her to willingly offer her voice to the discouraged musician. On the other hand, throughout the song's lyrics Miku is merely an observer, watching from within her virtual world and feeling pride, joy, and loss only in relation to the musician who composes words and melodies for her to sing. This passivity positions her as a tool, or a blank canvas onto which any number of software users can project their own emotions and ideas, through creative control of this idealized fictional girl. Even within the world of the music video, Miku has no capacity to act on her own and is carried about the room by the robot. The computer-generated singer remains boxed in by the four edges of the handheld screen, unable to interact with the band members with whom she is ostensibly performing and unable to cooperate with the robot seeking to help her. Even though her voice is passionately and sometimes almost breathlessly singing the lyrics of this song, our view of Miku is of a singer in a box, or rather, just the mouth of a singer in a box, with the rest of her body invisible and irrelevant.

By framing Miku in this way, the music video for "ODDS&ENDS" toys with some of the same questions that have been asked about past automation technologies. Similar to the ad that controversially called the drum machine "the drummer" on an Elton John record, and the concert review that framed a reproducing piano set onstage to perform in a concerto as taking the place of a pianist, "ODDS&ENDS" probes the nature of Hatsune Miku's function and identity in Supercell. Was Miku a digital tool, taken up as a substitute for a human singer out of convenience, and appropriately discarded when Supercell had achieved widespread success? Or could it be that Hatsune Miku, as a programmable vocalist, could actually be used by musicians to explore unique aspects of creativity that a human singer could not? As the lyrics in this song rehearse the criticisms with which Ryo had previously contended while using Vocaloid, the visuals of the music video suggest that Miku, too, may be just another piece of outdated, dying technology in a heap of discarded devices.

In the final scenes of the music video, as the musician in the song's lyrics cries out in regret at his inability to reanimate his creative partnership with Miku, the robot similarly fails in its frantic quest to keep her digital display powered. Suddenly, the driving rhythm guitar that had been propelling the emotional desperation of the scene cuts out completely, the handheld screen flashes a final warning, and Miku's image vanishes. The Vocaloid continues to sing over the scene, however, her disembodied voice continuing to narrate from somewhere beyond the cluttered room. Giving voice to the musician's regrets—and perhaps those of the toy robot—she sings, "I'm powerless, I can't even save a single one of these odds and ends" (僕は無力だ。ガラクタ一つだって救えやしない).

During this soft and vulnerable bridge, a change begins to occur in the room. Bolts and cogs slowly start to rise from the ground, and the tiny robot is lifted off its feet as well, gently turning head over heels as it joins a growing cloud of scraps gathering in the center of the room. A growing light, shining up from Miku's formerly blank handheld screen, begins to envelop the floating cluster of metallic odds and ends. Suddenly a brilliant flash lights the room, and a three-dimensional figure of Hatsune Miku materializes, her body given shape by the mass of odds and ends now suspended in the air (see figure 20).

The song drives to a climactic close as the camera pans around the hovering form of Miku. She slowly raises shining hands made of cogs and screws as the final lines of the song narrate a hopeful transformation:

FIGURE 20. Odd parts coalescing into the form of Hatsune Miku in the final moments of the "ODDS&ENDS" music video. Source: "【初音ミク】ODDS&ENDS—PV Full Ver.," *NicoNico*, posted August 12, 2012, by まさゆき, https://www.nicovideo.jp/watch/sm18592204.

> And at that very moment, the world's colors began to change.
> Sadness, happiness: it has all been experienced by this one person and these odds & ends.
> Lyrics become songs, and once again they pulse around the world for you.
> Entrust your intentions to that voice, now your emotions will resound.

The appearance of the life-sized image of Miku, composed of countless tiny odds and ends that had little significance on their own and previously lay discarded about the room, is a moving moment because of the way in which it evokes the collective efforts of the diverse community that built Vocaloid's popularity. It suggests that Hatsune Miku is not one single concept, but a shifting image that comprises countless small creative efforts. In the Vocaloid community, tiny contributions—from scrolling comments that caught on and created new trends, to uploaders offering up amateur choreography recorded on webcams in their bedrooms—were all part of the collective effort that gave rise to the Vocaloid phenomenon.

In the context of the song's lyrics, "these odds and ends" has an additional layer of meaning as Miku's label for herself, and the words suggest that a human musician using a Vocaloid is in fact a creative partnership. Earlier in the song, Miku asks of the musician, "The two of us came up with lots of words together, didn't we?" The relationship between humans and Vocaloid software is similar in many ways to the "man-machine" partnership that Kraftwerk articulated about their band and the machines they used. Like Kraftwerk and the techno-pop aesthetic that became popular in their wake, Vocaloid music became a separate genre of music not only because of the unique sound of its automated vocals, but also because of the way in which users collectively negotiated a distinct aesthetic for Vocaloid music that overtly explored themes of automation and machine-human collaboration, both in its lyrics and in its musical effects.

Examining the trajectory of the early years of Vocaloid music shows how amateur users developed Vocaloid technology along a vastly different course than its creators originally imagined. By doing so, an online participatory culture transformed a piece of niche software into a mainstream success on a scale that would have been inconceivable at the program's launch. Masataka Goto predicted in 2015 that Vocaloid's story was far from over and argued that singing synthesis would become as indispensable to the music industry as other synthesizers have.[33] In the years after Vocaloid's initial boom on NicoNico Douga in a primarily Japanese-speaking community, Vocaloid's reach became increasingly international, and Hatsune Miku's virtual identity raised new questions as her image moved into new spaces, including convention halls, classrooms, and live performance events in arenas of screaming fans in countries around the globe, which we will visit in chapter 6.

6. Holographic Performance
Automation and Participatory Fandom

On a hot summer night in 2018, I stood pressed into the middle of a cheering audience on the floor of a Manhattan concert hall, my merch-clad neighbors and I illuminated in a wash of neon teal light that shifted and pulsed with the rhythmic shakes of the glow sticks that the crowd was holding aloft. The performer onstage in front of me was Hatsune Miku herself, her swirling teal pigtails matching the hue of her fans' light sticks as she sang and danced to a high-energy two-hour setlist composed of some of the greatest hits produced by Vocaloid songwriters. On this night, however, my attention was not focused on Miku. I knew that her automated on-screen performance would play out identically in Mexico City next week and had unfolded word-for-word in Washington, D.C., two nights earlier. My interest was captured instead by the actions of her fans, who screamed and cheered for their favorite virtual star and spent two full hours executing an intricate choreography of carefully timed LED light stick moves and hoarsely shouted fan chants, demonstrating a precision with which an uninitiated observer could never hope to keep up.[1] I stood in the middle of the crowd, struggling to keep pace with the ever-shifting sea of bouncing and swaying light sticks, feeling both under-rehearsed and awed.

Just two songs into the show, and the audience's attention to detail was already on impressive display. The moment Miku concluded the opening chorus of hit song "Senbonzakura," the crowd, without hesitation, thrust their light sticks toward the ceiling and began shouting "Hey! Hey!" on the offbeats of every bar in time with their light sticks as the song plunged into an instrumental break. Precisely eight bars later, Miku began the second verse. The fans didn't miss a beat, dropping their chant and instantly switching their light-sticking to single emphatic thrusts on the downbeat. They didn't sing along; none of them uttered a single word. But if I had any doubts about whether they were tracking with the Japanese lyrics, those doubts were shattered when, seemingly out of nowhere, countless Hatsune Miku fans around me waited precisely until the end of the verse's fourth line to scream just the last four words together: "ONE, TWO, SAN, SHI!" The effect—of the unity, the passion, and the preparation embodied in those four words, shouted by the sea of fans surrounding me—raised goosebumps on my arms.

It is not every day that a fictional character headlines their own international concert tour, and it is no small matter that a character like Hatsune Miku—not a popular TV or movie character, but a fictional figure associated with a piece of Yamaha music software—reached a point in her global popularity where it became possible to draw thousands of concertgoers to sold-out shows in Asia, North America, and Europe. How did Hatsune Miku reach this point? The journey from Vocaloid's initial release in 2004 to this third North American tour in 2018 was gradual, and at times even experimental. Vocaloid continued to gain attention toward the end of the 2000s, and the new Vocaloid 2 engine, launched in 2007, made synthesized Vocaloid singing sound smoother and more expressive, contributing to the surge of online interest in the virtual singers. Starting in 2009, early efforts to bring Vocaloids into concert contexts were little more than brief experiments in which Hatsune Miku made short guest appearances on a screen in the venue, but by the mid-2010s, projection techniques had improved, and interest in Vocaloid concerts had increased as the virtual singers' online popularity had grown.

In the 2010s, concerts in which the digital performer's image was cast onto a transparent screen were referred to as "hologram" shows by fans and journalists, owing in part to the slightly translucent appearance of the projections. While the images were not truly three-dimensional, since they were projected onto a flat surface, the hologram name stuck and became associated with avatar performers of all kinds, including animated figures of musicians such as Whitney Houston and Roy Orbison.

Even as holographic Vocaloid concerts became more frequent events in the 2010s, Hatsune Miku and other Vocaloids became polarizing figures for music journalists in the West, with reviewers describing feelings of detached bemusement, disorienting sensory overload, and predictions of a dehumanized musical future. Many pointed to a lack of live interaction between performer and audience as a fatal shortcoming that would prevent holographic concerts from staking a permanent claim in the musical landscape. One reviewer argued that a 2016 performance lacked true communication and spontaneity, writing, "It's live in the sense that a throng of people watching a music video together would be. There's a crowd, but the show would be identical with or without their attendance. There's no interaction, no banter, nothing to encourage a unique experience. . . . There's a humanity lacking that makes it less of a concert and more of a public screening."[2] A columnist for the *Dallas Observer* in 2015 described Miku as "soulless" and "corporate synergy taken to its logical, creepy conclusion."[3] The reviewer argued that synthesized singing was "how the fall of human music-making starts," and—seemingly writing under the impression that these events were the first time in which automation technology had been deployed in a live performance—added, "I think Hatsune Miku, as a concept, is creepy as hell. . . . [B]ut I draw the line at holograms, because they're the first step in taking humans out of the music-making process altogether."

While it seems that this writer had little idea that his anxieties mapped directly onto the fears of other journalists who had sounded the alarm about player pianos and drum machines, and also failed to realize that artists had already been experimenting with "taking humans out of the music-making process" for many generations, he was not alone in his concerns. However, there was already a growing disconnect between journalists' assessment of Vocaloid as a lackluster holographic imitation of a live performance and the explosive worldwide growth of Vocaloid's fanbase in the 2010s. Hatsune Miku opened for Lady Gaga on her 2014 artRAVE world tour, was slated to perform at the canceled 2020 Coachella festival, and even made the jump to the classical music world, singing the lead role in an opera composed specifically for her, in more than half a dozen major concert halls worldwide. The growth of Vocaloid's popularity and influence suggested that there was more to these performances than risk-free replay and soulless singing.

Vocaloid's increased visibility in the music world at this time began to garner scholarly attention as well. Work by researchers including Keisuke Yamada and Masataka Goto examined Vocaloid's collaborative online culture, while Nina Sun Eidsheim and Yiyi Yin considered the processes through which the software's sounds and associated images contribute to the articulation of cultural values.[4] Questions about how profit and corporate interests affect Vocaloid fan culture have received consideration by Jørgensen and colleagues, who suggest that the relationship between Hatsune Miku and her fans "lingers between affective participation and corporate exploitation."[5] However, these types of concerns often stem from cultural differences about the management of copyrighted characters. While Crypton Future Media—the small Sapporo-based company that owns the rights to Miku's image—earns money from licensing deals for content such as video games, it does not receive royalties on the use of Miku's voice, and creators can use her image for any noncommercial purpose without needing to obtain permission. Although the company does produce and sell its own official merchandise, unofficial fan art and fan products are widely available online and at in-person fan events. The gray area around unofficial merchandise is particularly blurry in Japanese fan culture, where the unlicensed creation of fan art and other amateur products featuring copyrighted characters is especially widespread. In these subcultures, companies often ignore even for-profit copyright infringement, in part because of the value of the innovative contributions to the subcultures that these amateur artists make.[6] This chapter explores not only official Vocaloid performances created by Crypton, but fan-made events that use this creative flexibility to create their own Vocaloid concerts.

The majority of the scholarly work addressing Vocaloid to date has focused on the relationship between the music and its online fan culture, but little research has considered the culture and significance of live concerts. Including these performances is essential for a full understanding of Vocaloid's music and culture, but it also takes up a core issue that has been brought into focus across the past 120 years: the way in which live performance highlights the conceptual instability of automation. After generations of advances in musical automation technology, Vocaloid concerts have not brought us to a point at which performances are fully automated. These concerts still very much rely on nonautomated human creativity, even in places where full automation would be a possibility, and one of the most fascinating dynamics of Vocaloid performances is the way in which selective uses of automation technology as an artistic tool have facilitated and amplified unique forms of human creativity.

Taking Hatsune Miku's holographic career as an object of study enables us to question ideas of authenticity in live performances that incorporate automation and even to interrogate the dynamics of the performance event itself. Hatsune Miku's popularity is not a frightening step in the gradual sterilization of live musical performance; rather, it points to a larger creative trend that goes beyond a quest for clinical perfectionism. As holographic concerts became a fixture in the music world in the late 2010s, and virtual versions of performers including Gorillaz and Maria Callas were projected onto concert stages, Vocaloids' emerging context brought them into focus as one aspect of a much larger phenomenon. Examining these concerts as live performances poses certain unique challenges, because of the way in which a preprogrammed holographic concert seems to fall into a gray area somewhere between a human performance and a film screening. I argue, however, that the creation of cultural meaning and the most significant interactions occur not between the audience and a hologram onstage, but within the audience itself, and that this fan-led engagement is precisely the kind of creativity that is enabled in a live performance that involves automation.

The story of Vocaloid draws together examples of ways in which a new music technology facilitates musical creativity, first through the accessibility of the software itself, then through the collaborative creative process seen in the online community, and finally, in the performative audience engagement in live concert settings. In this story, I address the question of authenticity in live performance, not through contorted efforts to convince skeptics that a holographic Hatsune Miku can be equivalent to a human performer, but by showing how the audience's expanded role in these concerts confers a new type of cocreative agency. Although Vocaloid's detractors worry that an automated Hatsune Miku concert has been robbed of its liveliness and uniqueness, I show how the automated elements of these concerts are the very thing that enables the audience to play their own active and creative role.

In this chapter, I focus first on Vocaloid performances in North America and Europe, at concerts on Miku Expo tours in 2016 and 2018, in which the Vocaloid performer has been backed by a band of human instrumentalists in a format similar to a typical pop concert. I then contrast these performances with events in Japan at NicoNico Chokaigi and NicoNico Cho Party that forego key elements of human concert conventions and foreground the unique aesthetics and possibilities of a Vocaloid concert without relying on standard concert practices. In both case studies, I focus on aspects of the audience experience that have not yet been considered in the Vocaloid literature, such as coordinated audience actions and the detailed preparation fans undertake before the show. I include these concert practices in order to offer an updated context for understanding performance and liveness at Vocaloid and other preprogrammed concerts that feature participatory audience involvement.[7] When thousands of Vocaloid fans travel to live concert venues in order to express their fandom in a carefully rehearsed, embodied way, they create an in-person extension of the collaborative online Vocaloid community behind the music that is performed in these shows. This concert format is not a disconcerting replacement for performances by human singers, but the site of exciting forms of fan-led creativity and audience engagement.

VOCALOID FANS, JOURNALISTS, AND THE CONCEPT OF "REAL VERSUS FAKE"

The fact that popular Vocaloid videos have been viewed tens of millions of times leaves little room to question fans' enthusiasm for Vocaloid music, but some journalists have questioned whether this enthusiasm for a completely fictional singer is misplaced. In interviews conducted by media scholar Rafal Zaborowski, however, fans of Vocaloid music have explained clearly how the collaborative amateur production process behind favorite Vocaloid songs helps to create a deeper connection with the music. "It's just a bunch of talented people, using [the software] for everyone to enjoy," says one participant. "They don't have to go through [the industry], they just upload."[8] Although Miku's voice is computer generated, Vocaloid fans perceive her songs as possessing "real freedom of expression" that comes from the amateurs producing her songs. Participants described this in contrast with the hyperproduced images and music of major pop groups, which they described as "[a]ll fake."[9]

These values were called into question by reviewers outside the Vocaloid community when the characters first appeared in live concerts. Miku Expo, an international concert tour featuring Hatsune Miku and other Vocaloid characters, staged its first two US concerts in 2014.[10] However, even calling these performances "live concerts" has raised ire from music journalists. The great irony of Vocaloid criticism lies in the fact that detractors often disparage Vocaloid concerts by inverting this relationship between "real" and "fake" as described by fans. If "real freedom of expression" relies on a performer's ability to respond emotionally to real-time events, then a preprogrammed performance would indeed seem "all fake." The idea of freedom of expression in Vocaloid music, however, does not stem from spontaneous variations in live concerts, or even from the creativity of a single original songwriter who created the music. This sentiment arises from a sense of collective investment in the creation of Vocaloid works through online collaboration and layers of derivative works that interpret Vocaloid music in meaningful new ways, and it is also facilitated by the affordances of the automation technology that supports these endeavors. After a Vocaloid software user generates a sung vocal line, this unvarying automated performance becomes the locus around which other fans add to the unfolding meaning of a new piece of Vocaloid culture, complete with new creative elements such as animations or memes. It then enters the interpretive space of the live concert environment, where the automated singing still remains completely static and unchanging, but it is this very preprogrammed predictability that enables the audience to develop and perform their own role in the concert experience.

Miku Expo's more ambitious 2016 tour, which included nine shows in the United States as well as one concert in Canada and four in Mexico, garnered a significant amount of press attention compared to previous Vocaloid shows in the West, but much of it was negative. "If an ordinary show's affinity is to theater, Hatsune Miku's is to film: both are coldly automatic, nonspontaneous, even projected, with bleeding-edge invention, onto a vast and gleaming silver screen," wrote a Canadian reviewer. "The Miku Expo isn't a concert—it's a robo-show, a concert simulacrum. Two thousand fans hop and scream as before them light and sound take on the shape of real experience."[11] Another reviewer expressed boredom with a performance on the same tour, writing, "While Miku's performance was flawless, it was flat as hell. There was no story. She never lost her breath, or flubbed a dance move.... There was no payoff for me in the audience because there was no risk being taken by the performer."[12] Music journalism paints a picture of Vocaloid concerts as lifeless and uninspiring affairs—a depiction that stands in stark contrast to fans' experience of a vibrant and personally meaningful body of music. However, the friction between these contradictory reports highlights the places where the distinctive role of the Vocaloid audience is most visible, and where Vocaloid scholarship can make a significant contribution.

The relative absence of scholarship examining live Vocaloid events means that there is much to be gained from a consideration of the meaning of the audience participation at a holographic concert.[13] My approach to in-person interactions at Vocaloid concerts draws on Matthew Reason and Anja Mølle Lindelof's concept of "audiencing"—a process by which audience members bring a performance into existence with their attention and investment.[14] The concept of audiencing as a vector for feelings of liveness at in-person concert events means that ontological debates about liveness versus mediatization can recede into the background, and the focus can rest squarely on the relationship between the audience and the performance. This perspective is important for a study of Vocaloid concerts, because at these holographic events, the most important action is not on stage, but the actions with which the audience responds to the performance, and the ways that a live concert serves as a space in which fans gather to "enact the meaning of [their] fandom."[15] Much like the way in which the unchanging hole-punched notes of a player piano roll served as a predictable foundation for player pianists to craft a personalized performance, and the automated rhythm loop of a drum machine provided a consistent groove over which musicians could write new songs, the preprogrammed predictability of a Vocaloid performance is the very attribute that facilitates the audience's carefully prepared interactive contributions. In both the online Vocaloid community and the performance venue, fans' "creative co-production" and participation makes a meaningful contribution to the full experience of a Vocaloid concert.[16]

HOLOGRAPHIC BODIES AND HUMAN BODIES: FAN PARTICIPATION AT MIKU EXPO 2016

The projection technologies used at holographic concerts improved significantly after the first Vocaloid event in 2009, and performances in the 2010s created striking visual effects by casting the virtual singer's image onto a transparent band of plastic in such a way that she appeared as a three-dimensional, life-sized performer. Similar techniques have been used for other holographic performers, including the Tupac hologram that performed at Coachella in 2012 and the Roy Orbison and Maria Callas holograms that toured in the late 2010s. In concerts that use this technique, the screen is positioned front and center on the stage, occupying the same space as a human lead singer and creating the illusion that the hologram is singing and dancing independently. These concerts came to be known as "hologram concerts," referencing the three-dimensional images in science fiction franchises, but these performers' images do not actually move in three-dimensional space. With the transparent plastic enabling the audience to see the physical space around the singer, and the width of the setup enabling the hologram free rein of the stage, however, it's possible to suspend disbelief and occasionally forget that the figure on stage is purely a projection of light. Truly three-dimensional projections of performing musicians remain squarely in the realm of sci-fi futures, but the *holograms* descriptor stuck, and gradually journalists, the public, and even music scholars have accepted the term for use in describing concerts that feature a digital image of a musician on a screen in a concert setting.[17] The increasing use of these technologies for a widening spectrum of musicians that can be fictional, artistic, living, or dead suggests that performances featuring these digital avatars are more than a niche technological fad, and that artists are just beginning to explore the possibilities of this new performance medium.

While the technology itself is captivating, perhaps the most fascinating element of a Hatsune Miku concert is the dynamic between the virtual singer and her fans in a live setting. The audience responds to Miku as though she were a living person, chanting her name or screaming for encores, and the hologram gives the appearance of responding to their cries by wiping away a preprogrammed tear and thanking them before performing one more song. Cosplay—wearing costumes and accessories that represent a fictional character—is popular at Vocaloid events, and many fans who bear not even a passing resemblance to the pigtailed teenage girl arrive sporting teal-colored wigs or elaborate costumes. Most fans attend equipped with chemical glow sticks or LED light sticks, which are essential tools for participation at live Vocaloid events. These light sticks are not waved about haphazardly, however. A polished, unspoken set of guidelines governs the elaborate rituals of chanting and light stick movements executed at Vocaloid concerts. For the uninitiated attendee, the abrupt shifts between pulsing, double-time rhythms and fist-pumps accompanied by unified shouts at seemingly arbitrary moments can easily feel bewildering.

Although Vocaloid fans have put their own unique spin on light stick participation at live shows, the practice is common in East Asia, particularly at concerts featuring pop idols, where fans perform lengthy sets of elaborate dance moves, with customized routines specific to certain songs. This type of participation is a part of the reciprocal relationship of effort and dedication that fans and pop idols perform for each other.[18] Fans develop, share, and master these dance sequences in advance, and this style of synchronized group movement is also viewed as a type of cheerleading that shows support for the idol performing onstage.

Light stick practices quickly became an integral part of Vocaloid concerts in Japan, where Hatsune Miku and her Vocaloid counterparts are understood as "virtual idols." Although virtual idols are unable to see or appreciate a show of encouragement from their fans like a human pop idol would, these holographic idols still serve as points of collective focus for their crowds of admirers and reference points for the intricate light stick practices. When Vocaloid concerts reached Western audiences, the light stick moves came with them, but in these new contexts, not only was the object of light stick cheering (a living performer) missing, but the cultural context of these practices was absent as well, making the actions indeed as baffling—as one reviewer put it—as aircraft marshaling for invisible airplanes. The significance of these rituals is integral to the meaning of Vocaloid concerts, however, and the fact that there is no living performer at center stage helps to facilitate a closer study of the central interactions in this practice. Although there are other live concerts that include elements of coordinated audience participation, the presence of a human singer complicates the dynamics of these activities, since the audience's actions can be directed toward the performers or executed in order to get their attention. The fact that a Vocaloid character cannot see or hear what the audience is doing means that the crowd's activities are not attempts to communicate or connect with the performer, but actions that have meaning among fans.

As a relatively new performance phenomenon, these specific participatory actions and their meanings may also seem new and unique. However, Vocaloid performances have elements in common with other predictable participatory events such as cult film screenings, and these similarities are

Holographic Performance / 165

revealing as well. People attend screenings of *The Rocky Horror Picture Show* to enthusiastically fling rice and toast—and, in fact, even to shout calls and wave glow sticks—at the correct moments in the movie. Austin's classic study of cult film audiences demonstrated that it was the event itself (including the preparation, the waiting, and the active participation), rather than the film, that attracted and supported a dedicated fan following.[19] Similarly, Vocaloid fans attend live performances—and undergo the same rituals of preparation, waiting, and active participation—as an opportunity to participate in an interpretive community that is creating new cultural meaning in an audience of enthusiastic fans.

In addition to the preparation, waiting, and participation that contribute to Vocaloid fan performance, the actions fans undertake at a Vocaloid concert constitute a "ritual" because of the requirement of detailed preparation and the narrowly defined forms of correct participation. Fans who wish to participate must study the correct moves and calls—either online or at other Vocaloid events—until they are familiar with the timing, arrive at the event with a selection of colored chemical glow sticks or color-changing LED light sticks, and accurately execute embodied actions at the live event. Correct execution of the moves and calls is a marker of dedicated Vocaloid fandom, and performing these rituals in a group creates a sense of shared identity at the live concert event. As Clifford Geertz's work on rituals has shown, rituals themselves are performative acts that bring symbols to life.[20] While Vocaloid fan actions are similar to those found at pop idol concerts, the removal of the human singer from the stage does not render the audience's rituals meaningless, but rather turns the focus of the actions inward to the crowd itself, as meaningful collective signifiers of shared fan culture that bring Miku to life—not in an organic or physical way, but in a symbolic and imaginative way. Hatsune Miku's holographic image, then, is not a digital replacement for a human body, not a trick meant to dupe fans, but an animated symbol of Vocaloid culture and creativity.

FRAMING A HOLOGRAPHIC CONCERT

Miku Expo tours in the West were attempting to strike a difficult balance between tradition and innovation for their audiences during their early years. Vocaloid concerts in Europe and North America had been much less frequent than those in Asia, and the fanbase was neither as large nor as established. These tours offered an event that replicated many features of the Vocaloid concerts in Japan that committed fans had watched online, but they also needed to convey the appeal of a virtual idol to those who attended the concerts purely out of curiosity.[21] This was especially important given that the majority of journalists and concert reviewers who would be writing on the shows knew little of the Vocaloid world and would be a harder sell than the longtime fans, who would be instantly thrilled when the first few notes of a hit song from NicoNico rang out over the speakers. The format of these Western concerts needed to strike a careful balance between playing to the strengths of an automated concert and delivering a show that the audience would interpret as a "live experience."[22] For these reasons, Miku's North American and European performances on the Miku Expo tour were styled to resemble customary pop music concerts in many ways. At shows in Canada and the United States on the 2016 tour, the American band Anamanaguchi opened for Miku, providing a reassuringly conventional start to the evening by having humans perform onstage before the holograms. An opening band meant that the first thing the audience saw was a living musician, rather than a virtual spectacle. This opening act, however, was added only for certain concerts in the West and was not included for cities on the Japanese leg of the Miku Expo tour.

At North American shows, Hatsune Miku kicked off the Vocaloids' portion of the night with a rendition of "World is Mine"—a peppy pop track that has long been the most popular Vocaloid song on YouTube, with playful lyrics that do not mention Miku's virtual existence or her software, unlike some of the early NicoNico Douga hits. After this energetic opening, the Vocaloid cast delivered more than two hours of singing, dancing, and virtual costume changes. Concerts concluded with a lengthy encore in which Anamanaguchi joined Miku onstage for a pair of songs, then the lights dimmed, leaving Miku and a holographic Yamaha DX7 keyboard softly illuminated onstage. The virtual girl shyly sat down at the keyboard to accompany herself for one last song. Many found the shift jarring, with the attempt at the humanizing move rendering Miku even more inhuman, particularly when her holographic freedoms had been played up only moments before. Some audience members jolted by the sudden emotional change continued to cheer or laugh while others fiercely shushed them. The dissonance in this moment throws the fundamental issue in this performance into sharper relief: Was Hatsune Miku an automated simulacrum of a human, or a superhuman avatar who did not need to be constrained by the boundaries of reality? Instead of either committing to realism and believability as a flesh-and-blood performer in a pop concert or taking full advantage of the potential for superhuman spectacle, Hatsune Miku had slipped into an uncomfortable middle ground in which she was neither fully human nor fully hologram.

In his work on virtual liveness, Paul Sanden has argued that a piece of music that lacks a live performer can still be understood as a live performance when a "virtual performing persona" is invoked through a listener's encounter with highly mediatized music.[23] Even in the absence of qualities that are often used to describe live performances, such as spontaneity or interactivity, an audience perceives performative meaning when they are able to pinpoint an identifiable source from which the musical performance emanates. This identified source, however, does not need to correspond directly with a living performer. In a highly mediatized performance—or, even more so, in a virtual performance of a song for which no original human performance exists—individual listeners will have unique encounters with the music, the meaning of which depends heavily on their interest, education, and experience, all of which contribute to their concept of the performing persona.

If listeners' experiences engaging with a virtual performer at a Vocaloid concert are shaped by their relationship to Vocaloid music, then it comes as no surprise that journalists and fans walk away with different interpretations of the event. The original human source behind the vocal sounds produced by Hatsune Miku's singing voice are the samples created by voice actress Saki Fujita, who originally recorded every combination of phonemes in Miku's Vocaloid voice bank. Fujita is not the performing persona to which Miku is linked, however. Fans attach meaning to Hatsune Miku in two ways. On the one hand, the human source of Miku's music is the fans who dedicate their time to composing music for the Vocaloid, which expresses creators' human feelings, as Zaborowski's interviews have shown. On the other hand, in a Hatsune Miku concert, fans do construct a virtual persona for the Vocaloid singer and interact with her as a live performer, even though they are aware that there is no human singer onstage. The appearance of the holographic Hatsune Miku does represent the same character with which fans have interacted in videos, fan art, and amateur animations. This act of collective imagination is captured in Sanden's explanation of the way in which listeners cocreate a performing persona: "The persona is as much a character identified by any listener as it is one (per)formed by a composer or performer.... In this sense, listeners exercise as much agency in the creation of performing personae as do those responsible for the sounds on a particular recording."[24] Hatsune Miku, then, does exist—not as a living person, but through imaginative acts that draw on past experience and fan engagement to create meaningful interactions, brought to life by the process of audiencing.

In light of this concept of virtual liveness, how should an automated performer like Hatsune Miku be presented onstage? Some aspects of a Vocaloid's performance capabilities do exceed those of a human singer: Miku's range and stamina outstrip any living performer, and she can be programmed to pull off any new dance moves or superhuman stunts on a whim. In Miku Expo North America 2016, there were times when the performance took advantage of these larger-than-life abilities as an element in exciting visual effects including instantaneous costume changes and superhuman entrances and exits. For the most part, however, the concerts made a focused effort to frame her as just another human stage performer, giving just another concert that adhered to the standard format that audiences have come to expect. The difficulty in interpreting these performances lay primarily in the absence of a clear commitment to either a realistic or a spectacular production—at times, they clearly exceeded the bounds of a human musician's capabilities, reminding attendees that they were viewing something inhuman, but rather than fully committing to the idea of Miku as a superhuman performer, the shows remained tethered to the conventions of a typical pop concert.

Many critics have noted an important reality: when holographic concerts are compared side by side with pop music performances delivered by humans, Vocaloid concerts do come up short in important ways. First, while the projection technology used to place Miku on stage is impressive, issues such as light stick reflections on the screen and a loss of visual depth for viewers seated to the side of the stage can fracture the immersive experience for some, serving as a reminder that the audience is watching a projection and not an embodied performer. Second, while the concert's creators included preprogrammed efforts at audience interaction, such as having Miku introduce songs and express her thanks to the cheering crowd, Vocaloids are still unable to have a true dialogue with their audiences, eliminating the possibility for the unique glimpses into a performing musician's thoughts that can serve as memorable interludes in a concert program. Third, when a Vocaloid reaches a spectacular high note in performance, the awe listeners might feel if a human vocalist sang the same line is dampened, since the audience knows there was no physical challenge for the performer. Writers who have pointed to the absence of possible vocal flubs as a shortcoming of a Vocaloid concert are only scratching the surface of a larger issue. When we hear a daring rendition of a coloratura aria or witness a pair of trapeze artists at work, part of the thrill comes from the fact that the performers are pushing their bodies to the limits of human capabilities. There is inherent risk and a real possibility of disaster. Audiences are typically not hoping to witness failure, but the prospect of the performers missing a catch or a high A-flat adds emotional intensity because we empathize with the risk inherent in the high-stakes live performance moment.[25] The excitement of these experiences is lessened when we watch a prerecorded video, since we already know the outcome. With Vocaloid, too, there is no need to hold our breath—we know that the performer will always hit her notes flawlessly.

When we focus entirely on the losses and flatness of a performance by an automated singer, however, we lose sight of what we stand to gain from an exploration of this new performance format. Although Hatsune Miku's projected image resembles a human body, she is not a human performer. The most significant advantages of this substitution have little to do with her dazzling stage entrances, and although the preprogrammed voice and dance moves play out perfectly and without variation at each concert, this predictable perfection is not the most important accomplishment of this performance format. When audience members and critics focus on the strict repeatability of these concerts, they often fail to notice the site of the most significant, most fluid forms of participation and interaction. In fact, it is the very flatness and predictability of the automated performance medium that makes space for the dynamic fan participation that unfolds within the audience.

THE VIEW FROM THE FLOOR: MIKU EXPO 2018

After Miku Expo 2016, it was two years before a Vocaloid tour returned to the West. Miku Expo 2018 traveled across North America and, for the first time, extended its route to include Europe, with stops in Paris, Cologne, and London. Despite the additional geographical territory covered, Miku Expo 2018 received less press coverage than the 2016 tour. The sensationalized headlines and leads about inhuman pop stars were noticeably missing from newspapers and blogs as the tour made its way across North America. With news then circulating about upcoming holographic ABBA and Amy Winehouse concerts, descriptions of Hatsune Miku's automated concerts simply lacked the shock value that they had in 2016. Reviews from 2018 were published less frequently by mainstream media outlets and appeared mostly on private Vocaloid-, anime-, or tech-themed blogs. The tone of these reviews changed accordingly, from cynical journalists who focused on feelings of incomprehension, to fan bloggers' pragmatic considerations of the venue's acoustics and the content of the set list. Reviews in cultural publications with wider reach often changed their tone as well. Brandon Wetherbee closed his description of Miku Expo 2018's Washington, D.C., performance with the assessment that the fans were "enjoying a shared live experience" and noted, "The singer may not be made of flesh and blood, but this is an authentic experience."[26]

The emphasis on this event as a "shared" live experience is significant. Group interactivity—whether online through NicoNico comments, or offline in elaborate light stick moves—is an important part of Vocaloid music culture. As such, it is one thing to read reviews after an event, and it is another to be physically present at a concert as a participant. In a genre that is shaped not by the top-down dictates of a label but by fan uploads and fan interests, understanding the audience is essential to understanding live Vocaloid performances. It was for this reason that I purchased a ticket for Miku Expo 2018 in New York City and made the journey to attend this event and supplement the online Vocaloid discourse with research conducted at a live event.

Three hours before Miku Expo 2018 was scheduled to begin, I stood waiting in a line that stretched the length of two New York City blocks in the oppressive summer heat. Even as I stood in line, I saw that comments on the wide range of age groups represented at Vocaloid concerts were accurate, as were remarks on the enthusiasm of the fans. Hundreds of people filled a line that included preteens with their parents, twenty-somethings in elaborate costumes, forty-somethings carrying Vocaloid merchandise, and a handful of seniors sporting Hatsune Miku T-shirts. I spoke with serious fans who had flown in from out of state for their third Vocaloid concert in five years, and younger fans for whom this was their first live music event of any kind. The attendees toward the front of the line had set up camp with chairs and umbrellas early in the day in order to secure spots near the stage, and their eye-catching cluster of teal wigs and cosplay outfits was drawing stares and inquiries from passersby. Small clusters of people in line gathered around handheld speakers playing Vocaloid music and practiced light stick moves and chants quietly together.

Practicing in advance of a live concert—not as a performer, but as an audience member—is an important and relatively unique aspect of Vocaloid shows. At many musical performances, spontaneous audience involvement (such as waving lighters or phones) is commonplace, but fans have no need to dedicate time to rehearsing the motions and their precise timing within a particular song. Rehearsal is typically something that the performers onstage undertake, while the audience shows up to enjoy the performance with no prerequisite practice time. The fact that audience members prepare and rehearse for a Vocaloid show reveals that the fans themselves are engaging in an element of performance, even if they are not onstage.

This kind of preparation and engagement requires a high level of familiarity with the songs in the concert and the ability to anticipate the specific details of the music. Martin Barker has noted that music can "come a(live)" for audience members when they are primed for the experience and intimately familiar with the music being performed, describing an experience in which he was jolted out of his seat upon hearing his favorite moment in a

Mahler symphony.[27] The "experiential excess" Barker experienced depended less on the presence of traditional elements of concert liveness and more on his anticipation of the experience of that particular musical moment. Preparation, emotional investment, and participation are key elements of the fan experience at Vocaloid concerts as well. Repetition—knowing that a Vocaloid song will be performed without variation at a concert—enables fans to prepare by studying past performances and practicing the correct light stick moves, because they know that the artist will not perform an unexpected slow ballad arrangement of that song at the next concert. The predictable nature of these performances enables fans to create an independent layer of meaningful actions over a song that the audience then performs in live shows, rather than simply being passive observers. Concert attendees travel to the site of the performance in order to express their fandom through practiced participation that is performed in the crowd. In my observations at Miku Expo 2018 in New York City, these factors played an important part in the meaning of the show.

Pressed into the mass of elated fans on the floor of the concert venue, I watched as Hatsune Miku's energetic holographic dance moves were propelled by chest-thumping bass, blinding strobes, and a pop-heavy set list. I also watched the audience around me, in which dedicated Vocaloid fans spent two hours participating in coordinated light stick waving, carefully timed calls that they knew by heart, and cheers that grew hoarser as the night wore on (see figure 21). My experience watching these two aspects of the show afforded me the opportunity to consider the perspective of both the uninitiated journalist and the seasoned fan. Several of the songs I had previously studied while conducting research were performed at Miku Expo 2018. Since I was familiar enough with the music to be able to participate with the crowd in those moments, I understood the draw, the excitement, and the sense of togetherness that performing the moves and calls with the group could confer when I, like Barker, was primed for key musical moments. At other times during the concert, unfamiliar songs left me unable to participate at all. During these times, I found that just as journalists had written, I really did feel like I was merely standing and watching a music video in a crowd. I knew that the pixel-perfect renditions of those exact same dance moves had played out on transparent screens in other cities on the tour. From a certain perspective, repetition could seem to have been robbing the Vocaloids' performance of some of its significance, but from a fan perspective, I could see that this very same repetition and predictability was enabling complex group participation in the audience.

Compared with most pop concerts, at a Vocaloid show there are much narrower boundaries around correct and incorrect forms of participation. I arrived

FIGURE 21. Crowd of fans executing moves with light sticks during a song featuring Hatsune Miku at Miku Expo 2018 in New York City. Photo by author.

at Miku Expo equipped only with two green chemical glow sticks—Hatsune Miku's signature color, and the only glow stick color appropriate for cheering on the teal-haired Vocaloid—and this limitation prevented me from participating in songs that featured any Vocaloid other than Hatsune Miku, even if I had known the songs. During times like these, when I was merely watching the crowd around me instantaneously switching glow stick patterns at a climactic moment, I couldn't help but wish that I had been able to learn them myself, so that I could be in on the excitement. I considered the substantial amount of time that would have taken and also realized that for committed fans, that was entirely the point. Precision involvement at a concert like this is a rare opportunity for fans to express their fandom in a public setting—which, in the case of Vocaloid, is usually restricted to online interactions and private investment in activities like songwriting, video commenting, or purchasing collectibles.

A particular moment from the concert illustrates the distinction between repetition as loss and repetition as essential for participation. Partway through the show, after Hatsune Miku had sung several songs, there was a brief transition onstage, and Miku dissolved from the screen. As the audience cheered, the auditorium went dark, save for the glow of green light sticks still held high. There was a pause, and then a quiet, looping rhythm began to play out into the crowd. In these few moments, I did not know what was coming next, but the audience did, and I was able to watch the transformation take place around me. The crowd recognized this looping hook as the intro to the next song and realized that the next Vocaloid to appear was going to be Megurine Luka—the Vocaloid associated with the color pink—and began fumbling with the buttons on their light sticks, while exclaiming to their neighbors, "Luka! It's Luka!" Gradually, the sea of green light sticks began to shift to a pink hue, and the audience started up a rhythmic chant, shaking their light sticks in time toward the stage while the instrumental phrase continued to repeat. The anticipation escalated as the wash of light sticks grew more uniformly pink, and the chant grew louder. Then, in a sudden flash, Megurine Luka materialized onscreen and delivered the signature opening line of "Luka Luka Night Fever" with a theatrical fist pump, and the crowd lost their minds with delight. Coordinated light sticking failed for a moment as the audience screamed and cheered wildly. And I, standing in the middle of this outpouring of excitement, couldn't suppress a grin as I watched the sheer delight of the people around me, who were playing their own part in the performance, starting up the next appropriate rhythmic chant with smiles splitting their faces. The predictability of the details in Miku Expo concerts is exactly the type of repetition that draws criticism from reviewers. There was never an instance on the tour in which the audience got a version of this song that featured a different Vocaloid from the original, a fresh musical arrangement of the song, or even a slightly different set of dance moves from Luka, and this lack of variety has often been interpreted as a loss of spontaneity. However, this repetition was also the condition that enabled the audience to anticipate the appropriate forms of participation for this song and execute them as a group. Predictability was key to coordinated fan involvement in this moment in the concert, and rather than being a loss, it added a distinctive layer of engagement and group belonging.

As intricate as participatory fandom is at a Vocaloid concert, the complexity is not meant to prevent engagement or deter newcomers. The inclusive way in which Vocaloid enthusiasts seek to teach new fans the guidelines for participation is clear both online and in person, and this welcoming approach is a distinctive aspect of Vocaloid culture. Fan-made videos, as well as websites with written instructions and images, provide tutorials for the basic light stick moves used at a concert. At the venue in New York, attendees around me offered extra glow sticks to those who had come unprepared and took time to explain the colors. Perhaps most remarkably, a fan-made booklet circulated on social media during Miku Expo's run, containing thirty pages of detailed instructions for the moves and calls in the songs that were likely to be performed on the tour (see figures 22 and 23).

Doctor=Funk Beat
by nyanyannya ft. KAITO

oshiawase ni douzo

(Hai Chant)
mada aoi tori o osagashi no kata wa
Come on, now
akirame kirenai kata wa raise de
Get a chance
mou aoi tori ga mieru anata wa It's too late
genkaku ni mazohizumu ni yuuforia…
mousoushou mou doushiyou
kindanshoujou shiawase ga hoshii? Got it!

majiku? not majiku! tane mo shikake mo medikaru
kokoro mo karada mo zenbu hitatte choudai
oyobi kai? YES MY DOCTOR
oyobi kai? YES MY DOCTOR
kitai no kisai no sutekina kono DAITEN SAI!
Get on!

DOCTER FUNK BEAT toujou da Ready!
mei haiyuu soko wa ryoushin nan da itamu wakenai darou?
DOCTER FUNK BEAT go kagen wa Lady?
sono toori genjitsu wa kutsuu darake honto icchatte
happii happii oshiawase ni douzo

"sekaijuu no fukou niwa midasu no te o"
sonna anata niwa o kusuri o (Dumb down)
ima kono shunkan mo fukou ga habikoru
nara sekaijuu no gizensha wa itsu nemurun deshou ne?

rabu? near rabu! mousukoshi dake kemikaru
kokoro mo karada mo zenbu kuratte choudai
shiawase kai? YES MY DOCTOR
shiawase kai? YES MY DOCTOR
kiseki no isai no sutekina kono
DAI! DAI! DAITEN SAI!
Get on!

DOCTER FUNK BEAT kaishin da Ready?
Oh!! interijensu no mondai da naoru wakenai darou
DOCTER HELP ME kyuukan wa Rowdy!
koufukukan tarinai nara tanin o hataki otoshite
happii happii tsugi no kata douzo
(clap)

majiku? not majiku!
(magic? not magic!)
ruurie howan in? nai nai daa!
(ruurie howan in? nai nai daa)
oyobi kai? YES MY DOCTOR
oyobi kai? YES MY DOCTOR
ukatsu ni konya mo kyokudo ni sutekina kore ga
DAI! DAI! DAI! DAI! DAITEN SAI!
Get on!

dokutaa fanku biito yume o mita
dare mo ga shiawase ni naru kodomo
damashi no mousou show
dare mo ga fukouna no da sono fuzaketa disutopia no naka de wa

soko de koitsu da "kuraudonain" dashi oshimi wa nansensu
saa anata mo anata mo anata mo kore de yuki mashou
damuzeru in disutoresu youyaku anata no junban da Oh!! moushi okure mashita
moshimoshi shiawase ni naritai nara daremo ga tataeru sono na no tokoro e
sore ga dokutaa fanku biito
daiten sai DAITEN SAI!

oshiawase ni douzo

FIGURE 22.   Page from the fan-made callbook for Miku Expo 2018.

Miku Miku ni Shite Ageru ♪
by ika ft. Hatsune Miku

Intro:
(Kecha at slow part followed by Hai chant when instrumental picks up)

Verse 1:
kagaku no genkai o koete
watashi wa kitan da yo
negi wa tsuitenai kedo
dekireba hoshii na ano hi
takusan no naka kara sotto watashi dake eranda no
doushite datta ka wo itsuka kikitai na
(Se-No Hai Hai HiHiHiHi)

(PPPH)
ano ne, hayaku
pasokon ni irete yo
doushita no?
pakkeeji zutto mitsumeteru

Chorus:
kimi no koto
MIKU MIKU NI SHITE AGERU ♪
uta wa mada ne, ganbaru kara
kimi dake no watashi wo
taisetsu ni sodatete
hoshii kara
MIKU MIKU NI SHITE AGERU ♪
ichinenjuu kimi no koto wo
futari de uta wo tsukuru no yo
dakara chotto kakugo o shitete yo ne

(Hai chant)
"shite ageru kara"

Verse 2:
watashi no sugata mada me ni wa
mienai no wakatteru
dakedo watashi ikiteru kimi to hanashiteru
dakara,
buaachyaru no kakine wo koete afure
kaeru jouhou no naka
kimi to watashi futari de shinka
shite ikitai
(Se-No Hai Hai HiHiHiHi)

(PPPH)
yuube kiita
kimi no hanauta ga
ashita ni wa
watashi ga utaeru koto matteru

Chorus:
itsumademo
MIKU MIKU NI SHITE AGERU ♪
utatteku sore ga shiawase
tama ni machigacchau kedo
kizukanai furi wo shiteiru kimi wo
MIKU MIKU NI SHITE AGERU ♪
sekaijuu no dare, dare yori
kimi ni honki tsutaeru no
dakara zutto tonari ni isasete ne

(Hai chant)
donna toki demo

MIKU MIKU NI SHITEYAN YO ♪
saigo made ne, ganbaru kara
jishin wa aru keredo
sukoshi shinpai wo shiteiru kimi wo
MIKU MIKU NI SHITEYAN YO ♪
sekaijuu no doko ni ite mo
sagashi dashite tsutaeru no
dakara chotto
yudan wo shite agete

(song slows down; Kecha)
Miku Miku ni Shite Ageru ♪
mada mada watashi, ganbaru kara
kuchizusan de kureru
muchuu de ite kureru
kimi no koto

(Furi)
MIKU MIKU NI SHITE AGERU ♪
sekaijuu no dare, dare yori
daisuki

FIGURE 23. Page from the fan-made callbook for Miku Expo 2018.

The booklet's creators printed out hundreds of paper copies at their own expense, handed them out at the Los Angeles and San Jose performances, and sought volunteers on Twitter to print and distribute copies at other concerts. This callbook was not created for seasoned Vocaloid fans, but for people who were perhaps attending their first Vocaloid concert without detailed preparation and could benefit from explanations of what *furi* and *kecha* were, and exactly when to do them. The question of whether or not newcomers could actually reference these callbooks quickly enough for them to be helpful in a cramped and dark concert hall is another matter, but the generous intentions were apparent. It is important to note that this callbook was not prescriptive; it was merely descriptive, and serves as a record, in text form, of the practices that gradually coalesced around these songs over the years.

Although some of these practices originate organically in concert situations, many new additions have their origins online. Because of the interactive nature of NicoNico's commenting system, as discussed in chapter 5, comments that are shown in real time at catchy or climactic moments often snowball into "sing-along" moments, as growing numbers of users add their typed-out echoes to the comment stream. These practices have often carried over into live concerts, where thousands of fans who have never met but have watched these videos online can immediately coordinate their participation. The "missing story" that the journalists quoted earlier were looking for at Vocaloid concerts was not found onstage in the preprogrammed holographic dance moves. At Miku Expo 2018, the story was clearly unfolding all around me in this microcosm of participatory fandom. It was in the deep emotional connection the fans had with Vocaloid songs and the opportunity to join in with a practiced, embodied group expression of fan enthusiasm.

Miku Expo concerts offer Vocaloid fans a space in which they can collectively express their fandom, but by tethering Vocaloid performance to the conventions of human pop concerts, these events have opted to play it safe, remaining largely in the realm of simulation; in doing so, they have diluted the unique possibilities of this medium. In his study of robot opera—a genre of performances in which human presence is reduced or eliminated and technology is foregrounded—Alexander Sigman demonstrates how these technologically focused performances are not merely a gimmick that attempts to substitute human performance for machine performance, but they actually afford artists an expanded field of creative possibilities.[28] Sigman argues that these possibilities are facilitated by the way in which the boundaries between human and technology are blurred, enabling artists to experiment with new performance practices and explore questions of what it means to be human. In order to explore the expanded possibilities offered by Vocaloid performance, I needed to attend concerts given in formats that circumvented Miku Expo's reliance on pop concert conventions. When this scaffolding is stripped away, what new artistic opportunities emerge in Vocaloid performance, and what do they mean for creators and fans? The answers to these questions could not be found at Miku Expo, but in Japan, at events that have pushed the envelope of Vocaloid's creative potential beyond the pop concert template.

VOCALOID MUSIC AND AUDIENCE PARTICIPATION AT NICONICO CHOKAIGI

NicoNico Chokaigi, billed as an event that seeks "to recreate NicoNico's virtual world in real life," has been hosted by Dwango (parent company to Japan's largest social video site NicoNico) each spring in Chiba, Japan, since 2012. In 2019, Chokaigi—which translates to "super conference"—drew more than 150,000 visitors to the convention site in addition to 5.5 million live-stream viewers online.[29] After a pandemic hiatus, the 2022 edition of the event had 96,000 attendees, and in 2024, attendance had nearly rebounded to prepandemic numbers, with more than 125,000 guests visiting the convention site in person. Chokaigi events have featured exhibits and performances centering on the social video platform's most popular content—including VTubers, cooking, Kabuki theater, dance routines, video games, and Vocaloid—and offer fans the option of either purchasing a ticket to physically attend the event onsite in Chiba or paying to watch a live online broadcast of the performances featured at the event. In 2014, Dwango added NicoNico Cho Party to the annual calendar—an event that offered in-person and online tickets to a multinight series of performances at the Saitama Super Arena that focused on the types of music, dance, and video games that had become popular on NicoNico.[30]

Paid options for online viewership of in-person events are by no means unique to Chokaigi and Cho Party, but what is distinctive about these events is the way in which the organizers work to integrate the contributions of both the event's online viewers and in-person attendees. In April 2024 I spent two days at NicoNico Chokaigi, taking in performances, visiting booths, and observing both sides of the online and in-person participation dynamics at this event. Makuhari Messe, Japan's second-largest convention center, is located a forty-five-minute train ride from central Tokyo, and even before I spotted the sprawling buildings on the walk from the train station, I knew I was headed in the right direction when I saw fans sporting Hatsune

Miku hoodies, backpacks, and pins. Inside the convention halls, I circulated through the noise and crowds, pausing to watch an array of different performances on small stages throughout the venue. These performances featured many of the participatory creative categories of videos from NicoNico itself, such as "Tried Singing" and "Tried Dancing." At each of these stages, large screens displayed not only the usual close-up shots of the action onstage, but a live feed of the comments being contributed by online viewers who were watching the events in real time. These viewers' comments scrolled across the video on the stage-side screens, just as they do when watching any video on NicoNico. This system offered fans who were unable to attend in person an opportunity to participate from their home computers in a way that visibly added to the onsite experience. While other streaming services such as Twitch also feature live commenting capabilities during streaming events, Chokaigi and Cho Party integrate online viewers' contributions in a way that is visible to spectators standing in the convention center or arena, not only to other online viewers. This configuration dissolves part of the divide between "onsite" and "online" interactions, making the participation of an online attendee whose typed reaction appears on the stage-side screens just as visible in real time as an attendee who waves a light stick in the crowd.

Vocaloid events featured prominently at Chokaigi 2024, though they were not the preprogrammed hologram concerts seen at Miku Expo. The Vocaloid concert stage at Chokaigi featured the amateur producers of popular Vocaloid music, who performed live DJ sets that showcased their own NicoNico hits for enthusiastic crowds, who executed the fan chants and light stick moves with just as much dedication as at a hologram concert. As I watched a selection of DJs, the live comments that scrolled across the video feed, visible to both the online and in-person attendees, were frequently filled with expressions of admiration and gratitude for the producers of favorite songs. They also frequently mirrored the physical actions of the fans who gathered in front of the stage and participated with light sticks and chants, by typing out the chants or using characters to symbolize light sticks. DJ sets are not automated events like Miku Expo's singing holograms, and the DJs were able to take small liberties as they performed in real time. Nevertheless, the core elements that the audience was waiting for—moments for fan chants, for example—were kept carefully intact.

One of the most powerful moments of audience participation I witnessed at Chokaigi 2024 was during a DJ set by renowned Vocaloid producer 八王子P (Hachioji-P), who drew an overflow crowd to the Vocaloid stage area for his performance (see figure 24). Hachioji-P's set included a

Holographic Performance / 179

FIGURE 24. Stage at Hachioji P's NicoNico Chokaigi 2024 DJ set, with screen immediately above the stage showing comments from livestream viewers watching the performance remotely. Photo by author.

blend of newer and older songs, but when his 2016 hit "気まぐれメルシィ" ("Kimagure Mercy") began to play in the hall, the energy in the shoulder-to-shoulder audience shifted palpably—and so did the intensity of the participation. Although I had already stood in Miku Expo crowds that had performed the audience moves and calls mostly correctly, this committed crowd of Chokaigi Vocaloid fans delivered an audience performance that was stunning in its energy and precision. During the opening verse of "Kimagure Mercy," the fans light sticked with noticeably heightened intensity, but it was during the rhythmically propulsive chorus that the audience's contribution shone brightest. Miku's voice began to sing out the emphatic first line of the chorus's lyrics over the speakers: "*Zenzen atashi ni—*" and in the very next moment, the shouts of the audience were so loud that Miku's voice became almost inaudible, drowned out by the sudden call in which the crowd finished her line with her: "*—KYOUMINAI JAN!*"

This moment was the one in which I encountered Barker's concept of "experiential excess" most directly, amid a crowd of people who were experiencing the same thing as they anticipated the moment when they would shout that call together. Despite having attended much larger Vocaloid performances during my research travels, in the moment when this crowd suddenly shouted and raised their light sticks together, the electrifying intensity of the effect was unparalleled. At this small and simplistic concert, the tiny stage and unsophisticated lighting in the convention hall didn't matter. The oppressively hot concert area and sardine-can crowd didn't matter. Even the presence of Hachioji-P—fascinating as it was to catch a glimpse of the Vocaloid producer through the crowd, half hidden by the DJ rig—was not the driving force behind this moment. This moment was about the audience, and about the performance of a shared ritual in which a group of fans performed their part with all their energy in a unified expression of fandom. Taken together, the continuity of the fan practices at hologram concerts and convention-hall DJ sets shows that Vocaloid concerts are not merely automated replacements for traditional concerts, but examples of how Vocaloid fan practices themselves are a performance that has consistent features and meaning across different concert contexts, despite the fact that entirely different types of performances may be occurring onstage.

NICONICO CHO PARTY: FAN-MADE VIRTUAL CONCERTS

If a realistic hologram that simulates a lead singer's presence onstage is not a precondition for experiential excess, audience participation, or meaningful musical encounters at a Vocaloid concert, and if this genre of music is therefore not tied to traditional concert conventions, then what entirely new performance practices could this genre facilitate? Concerts at NicoNico Cho Party provide one possible answer. Though their names are slightly similar, NicoNico Chokaigi is a convention that includes small concerts in its event lineup, whereas NicoNico Cho Party is exclusively a nontouring concert series. These concerts take place at Saitama Super Arena, which boasts a 36,500-spectator capacity and facilitates a full range of concert sound, lighting, and special effects. Significantly, Cho Party Vocaloid animations are made by fans and are not official creations by Crypton Future Media, as are the Miku Expo concerts. These concert animations showcase the creative abilities of the amateur NicoNico community and take full advantage of the digital medium. With their intentionally foregrounded virtual aesthetic and played-up focus on the Vocaloids' computer-based nature, these performances belong to the same artistic lineage as the Kraftwerk performances that deemphasized the artists' humanity in favor of an aesthetic that focused and commented on automation. Here, however, unlike the jerky onstage motions of Kraftwerk that caricatured automation as a constrained and simplified version of human actions, the artists who animate these Vocaloid characters celebrate their digital origins and take full advantage of their technological possibilities. In other words, in the hands of a different generation of creative people with a different set of tools, automation has continued to serve as an artistic theme that invites exploration in performance, though one era's energetic digital pop stars bear no obligation to resemble another era's stiff-limbed robots.

Much like Chokaigi, Cho Party includes large screens adjacent to the stage to add a flood of online viewers' comments to the experience visible to the fans in the arena. During Vocaloid performances at Cho Party, the holographic avatars of Hatsune Miku and her friends are once again featured at center stage with very little visible human involvement, but introductions to the Vocaloid performances often begin not with an opening band, but with actors who present a brief skit in which they underscore rather than downplay the Vocaloids' virtual existence. In these skits, the actors introduce the Vocaloids by discovering them on a laptop screen or acting out a dramatic conjuring of the digital images. The Vocaloids themselves, by contrast, are never framed as human. Their virtual existence is foregrounded in these introductions as they smash their way out from behind virtual computer screens, materialize from masses of glitchy pixels, or feature in retrospective montages that recount Vocaloid's computer-based history, clearly emphasizing their digital context before they take the stage in any sort of performative fashion.

The first NicoNico Cho Party Vocaloid performance in 2014 offers a clear example of one such celebratory digital concert introduction and also serves to shed light on contrasts between a Cho Party performance and a Miku Expo performance. Video footage from the 2014 Cho Party concert shows four human performers standing at the front of the stage around a nineteenth-century-style street performer's wooden barrel organ on a wheeled cart. One actor turns the crank on the organ slowly and intentionally, and the actors all sway in time to a music-box tune that plays out into the packed concert venue. The mechanical notes of the music box melody chime steadily into the quiet of the arena for a few moments, and then eventually halt, but the human organ grinder continues to rotate the crank in a slow, seemingly ineffectual circle. For what purpose? This scene onstage frames the concert, and even Vocaloid music itself, in a significant way. The barrel organ is a preprogrammed musical machine, similar in many ways to Vocaloid, with pins on a cylinder in place of pixels on a screen. Linking the barrel organ with music at a Vocaloid concert serves to reinforce the holographic singers' image as automated performers. This visual connection positions Vocaloids as descendants of the barrel organ, the player piano, and the sequencer in the history of automated music making and sets up the context for the performance the audience is about to experience.

As the organ grinder continues to rotate the crank, instead of another barrel organ melody, the disembodied voice of Hatsune Miku abruptly fills the arena, spitting out a rapid-fire string of monotone syllables at a speed so fast as to be all but incomprehensible. The spotlight gradually fades on the organ grinder—who is still dutifully turning the crank that now seems to be powering the inhuman stream of syllables uttered by the automated Vocaloid—and as Miku's outline gradually becomes recognizable on the screen, an arena full of fans scream their approval and raise their light sticks. The sea of multicolored light sticks in the arena gradually shifts to a more unified wash of Hatsune Miku's particular shade of green as fans identify the Vocaloid's silhouette and click their devices to chromatically align themselves with her. Still emitting a droning stream of uninterrupted syllables as she finishes materializing onscreen, Miku takes up her usual series of pop-inspired dance moves, but while her body is moving at a comfortable pace, in an eerie juxtaposition, her mouth continues to work at a superhuman speed that almost becomes a blur. Here, any illusion of humanity akin to Miku Expo has already been shattered.

The title of the NicoNico hit that opened this 2014 concert is "初音ミクの消失-DEAD END-" (*Hatsune Miku no Shoushitsu -DEAD END-*, "The Disappearance of Hatsune Miku -DEAD END-").[31] Its infamous lyrics are spit out at an uncanny rate of 720 syllables per minute in the verses—a speed thought to be impossible for humans (though this hasn't stopped users on NicoNico from trying!).[32] The song is a first-person narration by Hatsune Miku, describing her feelings of despair and her examination of her unstable identity, as her software glitches and eventually shuts down. Certain moments in the song utilize heavy vocal distortion to achieve effects simulating digital decay as Miku gradually fades from existence, drawing on the effect of glitch aesthetics to suggest a breakdown of communication and anxieties about the failure of digital systems.[33] In the final seconds of the song, after the Vocaloid haltingly utters a shaky goodbye, a message repeats: "A critical error has occurred. A critical error . . ." (深刻なエラーが発生しました).

Opening a concert with a grim underscoring of the limitations of the singer's purely digital existence does not in any way serve to generate the organic rapport of a concert given by a human performer, and its glitchy and monotone delivery would in no way reassure the Western critics of the concerts who felt as though they were watching a robotic music video.

FIGURE 25. Hatsune Miku intones "I want to sing!" as warning messages tile over the screen during a performance of "The Disappearance of Hatsune Miku" at NicoNico Cho Party 2014.

Rather than creating an approximation of a human performance, this concert opener takes the opposite approach, emphasizing and exploring Miku's inhuman qualities. The Cho Party animation for this song does not anchor Miku to a particular physical space or a set of realistic stage props, nor does it even maintain Miku at a consistent scale that preserves a sense of the spatial relationship between the audience and the stage.

Textural effects like cracks and error messages appear between the Vocaloid and the audience, reinforcing the viewers' awareness of the screen's mediation. Lyrics that could have been unobtrusive subtitles appear over and around the Vocaloid's body in a range of striking typefaces, sometimes even colliding forcefully with her virtual body, knocking her off balance. Forced perspective shifts utilize techniques such as extreme foreshortening to create the appearance of Miku rising into the air or stretching a hand out into the crowd (see figure 25). NicoNico Cho Party performances are stripped of many of the customs of concerts with live singers. There are no opening acts, no band introductions, no live backing musicians, and no encores. Much more so than other concerts and tours, these performances take advantage of the virtual nature of Vocaloid singers, as the fan animators throw the characters' images about during numerous instantaneous set changes, drop them underwater, or smash them through glass walls—all while they continue to sing without drawing a single breath out of place. The effect is anything but human. Far from being a shortcoming, however, this approach highlights the artistic possibilities of the Vocaloid medium, amplifying and celebrating the digital freedoms that a nonhuman singer affords.

CONTRASTING CONCERT FORMATS

By adhering to the conventions of human concerts, touring events such as Miku Expo ask their audiences to consider a Vocaloid concert on the same terms as a performance by any other pop star, and reviewers rightfully treat them as such. For many of these critics, however, an essential part of a live concert is the physical presence of a performer who is able to create a feeling of organic rapport with their audience. Detractors who pointed to holographic singers' inability to interact with a crowd were correct—the Vocaloids at these concerts possessed no capacity to respond to their audiences in real time. There was no human connection between Hatsune Miku and her fans over the course of any of these Vocaloid shows. However, while tours in the West have frequently attempted to downplay and paint over this perceived lack with humanizing elements such as acoustic encores, the presence of human backing bands, and an assortment of responses and comments to the audience preprogrammed into the show, NicoNico Cho Party opted for an opposite approach in many regards, choosing to foreground and celebrate the holographic artifice to which touring concerts generally avoid drawing attention. At these events—and at any live Vocaloid performance, including those in the West—the relationship between the figures onstage and the crowd matters, but it is less important than the relationships that already exist within the community and have been built around the virtual figure of Hatsune Miku through the efforts of creators who compose and collaborate, and fans who participate in the creative culture that gave rise to the Vocaloid phenomenon. Being present at a Vocaloid performance is about being part of a collective experience, in which the performance of fan rituals confers a sense of belonging and a heightened emotional experience. Every time a new Vocaloid's voice plays over the sound system, and the excited crowd switches the color of their light sticks, the show of color isn't directed at the Vocaloid performer. It is a participatory act that connects the light stick bearers through a sense of belonging and unity with the rest of the crowd—an insider's badge that identifies the participant as a member of the same amateur community that views, comments on, and connects this diverse web of creative contributions.

Taken together, it is the events that run on the creativity and dedication of fans that have taken the biggest artistic risks and that have best explored the Vocaloid fandom's potential for participatory interaction. In addition to the concerts visited in this chapter, many other fan-run events with Vocaloid-themed DJ sets or simple projection-screen setups have taken place in cities including London, San Jose, Copenhagen, Toronto, and Berlin, where people have gathered to enthusiastically participate in Vocaloid rituals and enjoy musical experiences with other Vocaloid fans. One such event organizer, Tito, articulated his feelings about the meaning of the shared experience at a Vocaloid concert—even one like his small gathering, which used little more than a small screen in a blacked-out university classroom:

> That's the whole joy, and the whole reason why people would go to these hologram concerts, right? Like, people ask, "why do you wanna go to watch a virtual person, or a virtual singer?" Like, it doesn't make sense, it's not real, it's not a real person, right? But it's the feelings and emotions and the fact that you know, we're all there together, who have shared, or have memories, or have had tough times in our life that have been helped by the songs from this virtual singer, or the community around this virtual singer, you know? It's all of those things, it's all of the feelings and emotions that really is the fruit of the whole experience.[34]

In any Vocaloid concert, the key actors are the countless passionate contributors who, since 2007, have devoted their time to songwriting, arranging, Vocaloid tuning, adaptation, dance creation, freeware development, animation, online commentary, motion capture, live concert participation, and fan event organization. The reviewers who attended Miku Expo concerts looking for spontaneous improvisation by expert musicians were attending the wrong type of concert for the specific experiences they sought. However, the story of Vocaloid fan culture shows that automation is not simply a simplistic substitute for spontaneity, and holographic technology is not something to be feared as a blanket replacement for human presence. In order to create a setting in which enthusiastic preparation and participation from fans can be a significant part of the show, two factors must be present: predictability and repeatability. Automation permits repeatability, and repeatability permits deeper investment in participatory fan culture. Our era of digital communication has enabled performative fandom to become more pervasive and more precise than ever before. Looking past the initial strangeness of Vocaloid concerts, these concert formats offer important insights into the nature of audience engagement in the twenty-first century. At events like these, the acts of fandom being performed are not directed toward a human performer. In many of the cases discussed in this chapter, there is no human performer present, and fans are not screaming for the attention of a figure onstage. The performance taking place in the audience is directed inward, as a performative act that is both an expression of personal fandom and a participatory action directed toward other fans. Participation, rather than passive observation, is key to the creation of new meaning in this environment, both online and at concert events that test the boundaries of our definitions of live performance.

Conclusion

In a May 2024 promotional video for Apple's iPad Pro, a close-up shot frames a ticking metronome in a dimly lit space. A quick cut jumps to a record player, which pops and whirrs softly before the opening chords of Sonny and Cher's "All I Need Is You" warble unsteadily to life. After this intimate prelude, a wide-angle shot reveals a spacious room as it floods with light, revealing an elaborately piled display comprised primarily of musical instruments, art supplies, and games—as well as a massive hydraulic press, the upper plate of which is poised ominously above the colorful stacks.

An abrupt industrial clunk signals the start of the press's descent. A trumpet is its first casualty, and a melancholy horn sound weaves in with Sonny's vocals and the crackling metal as the instrument twists and crumples beneath the press. A vintage arcade cabinet is next; sparks burst from behind a panel and the game emits a descending 8-bit "failure" motif as the display switches to a GAME OVER screen. Is the viewer meant to wince, cheer, or merely be enthralled by the carnage? The camera cuts to an upright piano with cans of colored paint on top, and we can anticipate its fate well before the bursting cans drown the collapsing keyboard in turquoise and harvest gold. A poseable artist's mannequin folds beneath the press with its hands upturned in a silent plea against its impending demise, and an alarm clock begins to ring while the increasingly dense contents of the display crunch, snap, and shatter in a cacophony of obliteration. The hydraulic press reaches the end of its path, and the scene falls silent for a moment, before the plate lifts to dramatically reveal Apple's thinnest-ever iPad. As the ad comes to an end and the final shot showcases the glossy, slim device, Sonny's voice croons in closing, "All I ever need is you."

The intended message behind Apple's ad, which was titled "Crush!," is readily apparent: all of these instruments, games, and tools fit on a single lightweight device that can facilitate creativity and fun. This was not the message that many viewers took away from this ad, however. Within forty-eight hours of its premiere, countless creative professionals—from musicians and film directors to authors and actors—had taken to blogs, social media posts, podcasts, and YouTube videos to articulate how disturbed they had felt while watching the one-minute video. Even outside of the arts world, tech industry writers panned the advertisement as "disgusting," "heartbreaking," and "insensitive."[1] The collective outcry that flooded the internet was so intense that Apple issued a rare apology and retracted "Crush!" from TV. "We missed the mark with this video, and we're sorry," said Tor Myhren, Apple's vice president of marketing, just two days after the ad was first shown.[2]

Videos featuring objects being destroyed in detailed slow motion were by no means unfamiliar or unwelcome sights in 2024. The Hydraulic Press Channel reached ten million total subscribers on YouTube in December of the same year, and the channel had even featured the pressing of some of the very items that were demolished in "Crush!," with books, drums, paint cans, television sets, and guitars common to both.[3] In the context of the twenty-first century's fascination with slow-motion destruction, Apple's ad could have been a hit. So why was this ad the target of such intense backlash, and why did it feel so unsettling for so many?

The horror of "Crush!" had less to do with the carefully orchestrated visuals and more to do with the fact that the ad unwittingly functioned on a symbolic level, enacting some of our worst cultural fears about the relationship between technology and art. The video asked viewers to watch as tools that have long been used for creative human pursuits were destroyed by an unstoppable and impersonal force, and it concluded with the implied suggestion that "all [we] ever need" is a shiny new device. Many commentators noted that 2024 was a particularly disastrous cultural moment for Apple to run this ad, owing to the intensifying anxieties about large technology companies' relationship with the arts during the preceding year. The use of AI as a replacement for screenwriters had been a key issue in the lengthy strike by the Writers Guild of America in 2023. A song that used new generative AI tools to fraudulently mimic the voices of Drake and The Weeknd—and did so passably enough to accrue millions of streaming plays before being taken down—raised serious questions about fraud, intellectual property, and artist compensation in a changing technological landscape. In light of these and many other stories about this tense relationship, Peter C. Baker asked of Apple's "Crush!" ad, "Did no one point out that people are increasingly wary of tech companies' impact on the creative professions?

That people have soured on Silicon Valley's apparent desire to monetize human creativity in as many ways as possible, from extractive streaming arrangements to harvesting human-made art as A.I.-training material?"[4]

If we distill both sides of the story of "Crush!" down to statements about values, Apple was aiming to depict the *concentration* of value—a story about the new iPad Pro as a device that could facilitate many creative pursuits—but in the context of heightened technological anxiety, viewers saw instead the *destruction* of value—a story about the new iPad Pro as a device that would eliminate tools with long histories of specialized and skillful creative use. Tools that had been valued for generations were framed as discardable, and the salt in the wound was the way in which their destruction was seemingly glorified. This advertisement and the resultant outcry made one thing extremely clear: the anxieties surrounding technological takeovers and inauthenticity that have been explored in this book are, to say the least, neither resolved nor extinct.

The dialogue around "Crush!" brings us back full circle to the advertisements from chapter 1, the types of stories they told about player pianos, and the criticism and distrust expressed by their detractors. When critics of Apple's iPad ad pointed to the instruments and art supplies that were pressed as symbols of hard-won human artistic accomplishment and decried their substitution with easy-to-use apps, they mirrored the very same indignation that was voiced by critics of the player piano, who felt that the technical shortcut provided by its mechanical fingers robbed music lovers of the value of slow and labor-intensive piano study. Similarly, the discomfort about the *Billboard* "Drummer on the Hit Record" ad for the LM-1 examined in chapter 4 prompted controversy because it, too, seemed to suggest that the machine could replace a skilled musician trained on acoustic drums. This recurring backlash does more than reveal a lineage of unpopular advertising tactics, however; it illuminates important insights into our most closely held values.

Consider the illustrative difference between "Crush!" and a hypothetical ad from a hypothetical technology company that depicted a hydraulic press destroying everyday irritants—perhaps utility bills? dirty dishes? scam calls and emails?—and how this would have prompted a completely different reaction from viewers. Intense emotional responses to ads like "Crush!" are valuable sources of insight because they highlight the things we are most afraid to lose and even underscore aspects of what we believe it means to be human in a technological age. Controversial advertisements push up against "sore spots" on our collective awareness in a way that can be challenging but can also contribute to constructive dialogue. By revealing the things that we do not want, technological anxiety helps us to better understand the things we want most.

WHAT WE FEAR, AND WHY WE FEAR IT

The anxieties about automation technology in the arts, as explored in this book, can be categorized into two broad areas:

1. The fear of a loss of authenticity
2. The fear of a harmful redistribution of power

Concepts of authenticity are highly subjective, constantly shifting targets that are dependent on cultural context—as several stories in this book have highlighted—but despite the instability of authenticity as a concept, fears about its potential loss are rooted in a genuine desire to preserve quality and integrity. This fear typically plays out in the context and customs of a musical genre or long-standing musical tradition. Critics who worried that people would be less inclined to spend time on piano lessons if the player piano afforded an easier alternative revealed a belief in the value of hard-won musical skills. Mel Lewis's anxiety that rock drummers would be "wiped out" revealed a belief in a traditional form of drum performance that is rooted in real-time collaboration and creative flexibility. Music critics' fear that Vocaloid would replace live human performers revealed a belief in the value of embodied, responsive performances by vocalists who embrace risk to foster human connection with their audiences. These types of anxieties boil down to the fear that something we value deeply—especially if it is costly or difficult to attain—may no longer be valued by others. It is a fear of being left behind if a critical mass of people abandon our own closely held beliefs in favor of something new and different. It is not, however, a fear of the technological tool itself.

A loss of authenticity in a small group of people's musical practices is in itself not an overwhelmingly significant change, but when the fear of a harmful redistribution of power enters the picture, the stakes are higher. When power changes hands, it is not into the hands of machines, however, but into the hands of people who own and manage the machines. It is these people—not the machines themselves—whose goals and values can have significant repercussions, especially as these values become crystallized in the interfaces, laws, and practices related to new technologies as they develop. In twenty-first-century life, this fear has become more pronounced as we feel inundated by increasing amounts of unavoidable ads, dissected by invasive tracking, and bled dry by soaring subscription costs. The cynicism that results when we feel exploited by corporations who have new technological tools at their disposal and are wielding them in exploitative ways is building. Societally, we are indeed facing a significant issue, though it is not with sentient pieces of automation technology plotting to overthrow humans, but with a clash of values.

Fears about the redistribution of power do not just include anxiety about CEOs or record label executives controlling and weaponizing technological tools, however. The fear often extends in both directions and reveals itself in anxiety about the implications of a redistribution of power to amateurs who might lack the experience and artistic sensibility to discern quality art. John Philip Sousa feared that putting player pianos and phonographs into the hands of low-brow listeners would erode the national musical taste and result in Americans accepting mechanically reproduced music in place of live performances. In discussions about the accessibility of online streaming platforms and the proliferation of "bedroom producers" who create music independently in home studio setups (which include, but are certainly not limited to, Vocaloid producers), critics often bemoan what they perceive to be the vulgarization of the music industry, and the flooding of the market with low-quality music that is the result of a democratized digital recording process putting creative power in the hands of less-skilled creators.

The fact that this fear extends in both directions and reveals itself in anxiety about the consolidation as well as the democratization of power reveals that there is no single answer to the question of who should have power. An industry in which power and resources are centralized in the hands of a few influential record labels has the ability to fund albums with large recording budgets and employ more musicians, but recording artists' stories of disempowerment and struggles for fair treatment show the shortcomings of this approach. On the flip side, in the twenty-first century, the democratization of the music industry through affordable recording equipment and the growth of artists' ability to upload and promote their own music have resulted in a musical ecosystem in which more than one hundred thousand brand-new tracks of music were uploaded to Spotify every day as of 2022—a total that is more than the number of new tracks created during the entire calendar year in 1989—and new artists struggle to differentiate themselves in an oversaturated streaming market. Musicians have difficulties in both scenarios, but divided fears about the abuse of power higher up in the industry and fear that "the masses" will abandon quality art leave us with a messy and indefensible middle, suspicious of those above and below, in which different groups of people have different ideas about which sliver of the music community should hold the power to determine whose art is valid.

192 / Conclusion

WHERE DO WE GO FROM HERE? DEVELOPMENTS AND AFFORDANCES

With this much fear and resentment swirling around the development of new automation technologies in music, it might seem that the creative world could be better off if we turned back the clock or slammed the door on new technologies in order to protect what we have. What exactly does the ongoing relationship between automation technology and music actually afford us? Melding automation and music facilitates two types of developments, which can be broadly categorized as follows:

1. Innovation: Folding new automation technologies into our creative practices facilitates truly boundary-pushing art that takes new sounds, techniques, or tools and forges new creative practices that exploit technological possibilities.
2. Democratization: Introducing new automated tools enables groups of people who would previously have been denied entry to certain parts of the world of music—either as listeners or as creators—to access and make art more easily or affordably.

It is essential to understand that these developments are in fact two sides of the same coin. The stories in *Automatic Artistry* illuminate the ways in which innovation and democratization play out in tandem. The affordances of the player piano enabled composers such as Nancarrow and Stravinsky to realize musical ideas that were so complex and precise that they were physically impossible to execute without this technology—whether rendered by an expert player pianist like Rex Lawson or programmed into a reproducing piano roll for exacting control. At the same time, music lovers who owned a player piano were newly able to study, enjoy, and engage with music they may well have never otherwise heard in their lifetimes. Even though most of these amateur player pianists' renditions were of a much lower musical standard, the lowering of barriers to access also facilitated encounters with music that would have previously been "locked up" from most music lovers, as Bertram Smith articulated so well in 1911.

In the world of synthesizers and drum machines, Kraftwerk and YMO developed a creative dialogue between humans and machines in order to comment on the role of technology in modern life. Elsewhere, owning a drum machine permitted songwriters to work in the privacy of their homes rather than paying for costly studio time, just as Roger Linn imagined. In the decades since, an array of creative inventors and musicians have developed entirely new genres of music and performance traditions, including

EDM and hip-hop, which feature synthesized timbres and beats unique to their respective genres. These new streams of musical creativity now boast their own Grammy award categories and a legacy of artistic innovation that has birthed significant creative trends in music.

In 2013 Hatsune Miku starred in an opera called THE END, not as a replacement for a human soprano, but in a role written specifically for her by composer Keiichi Shibuya. THE END—which was performed in major concert halls in Europe, Asia, and Australia between 2013 and 2018 and featured costume design by Louis Vuitton—used Hatsune Miku's imagined virtual existence to interrogate ideas about death, identity, and artificiality in a way that maximized the potential of Vocaloid's unique lens to striking artistic effect. Amateur songwriters continued to use Vocaloid to participate in a unique musical subculture, and in Japan, a simplified educational version of the software was introduced in classrooms as a tool to help children develop basic songwriting skills. Hatsune Miku's vocals have been used in a range of innovative and democratized applications outside of the Vocaloid subculture. Sophisticated video game soundtracks use the attributes of her synthesized voice to connect to themes in the plot, while other creators make humorously absurd use of her voice, such as the high-pitched, warp-speed vocals of "Nyan Cat!," the fifth-most-viewed YouTube video in 2011. In every story of automation technology in music, innovation and democratization go hand in hand.

"But wait," some might say, "is it worth it? Have EDM and colorful cat memes been worth the sheer upheaval that we have wrought with our new technological tools?" I will be the first to concur that merely pointing out the potential benefits of automation technology is not argument enough to brush aside the concerns of more than a century of recurring anxieties. However, by verifying first that it is not the technological tools themselves that people really fear, and also that these technologies have facilitated an array of new and exciting forms of musical creativity, the challenging questions that remain become easier to confront, though confront them we must. At its heart, anxiety about automation in music is heightened by perceived differences in values; it is the fear that other people will not value the things we value, and that widespread change will leave us in the dust, clinging to the remains of ideas and activities that we believe are essential to the human enterprise. This fear arises in connection with cultural change of many kinds, not just in music. However, these values we hope to protect have less to do with whether someone creates a melody by programming a synthesizer or playing a saxophone, and more to do with our concepts of hard-won skill, human connection, integrity, justice, and creativity. The discourse surrounding new technologies has a particularly incisive way of revealing our cultural values, because of the way in which their application can test the boundaries of our existing perceptions and practices. But if the loud and reductive fuss about technological takeovers is only a red herring, then the values behind it still need to be properly excavated and understood.

This is the hidden benefit of technological anxiety—a third development, if we choose to engage with it: clarity. Taken together, these fears shine a revealing light on our ongoing efforts to negotiate our values. New technologies in music often serve as "edge cases" that test the boundaries of our definitions and practices. Are you a musician if you pedal a player piano? Is a drum machine an instrument? Can a virtual singer give a concert? These questions trouble our comfortable classifications, and that is why they are more fraught, like skirmishes at the boundaries between two territories, the resolution of which can shift our existing concepts and have significant cultural ramifications. Additionally, and very significantly, technological anxiety also highlights new ethical issues that arise alongside these blurry boundaries, focusing a spotlight on the potential gaps in practices and policies in newly emerging gray areas where artists' rights have not yet been clearly defined.

As one of the most vitriolic opponents of player piano and phonograph technologies, John Philip Sousa's warnings in 1906 about the "menace of mechanical music" portending the downfall of amateur musicianship and good taste seem in retrospect to be overwrought alarmism, but the composer's forceful comments take on a different meaning in their full context. As Patrick Warfield has illustrated, Sousa's 1906 article was a tactical maneuver to sway public sentiment in order to gain support for an amendment to copyright law that protected composers' rights.[5] The bandleader's concern had less to do with the shrinking of the "national chest" and more to do with the fact that mechanically reproduced music—unlike concert tickets and sheet music sales—was not regulated by existing laws, and composers received no royalties on the rapidly proliferating piano rolls and phonograph recordings that contained their work. In "The Menace of Mechanical Music," after his series of rhetorically charged vignettes that warn of the impending downfall of American music, Sousa turns to the real matter at hand: "why the powerful corporations controlling these playing and talking machines are so totally blind to the moral and ethical questions involved." There, at last, is the real crux of the issue, in Sousa's own words. The technological fear stems from questions about who has power, what values they are reinforcing, and whose interests they are serving. The debate around player pianos and phonographs was not entirely about technological substitution but about just economic practices and composers' rights to earn income on the reproduction of their pieces.

In 1909, thanks to the efforts of Sousa and others, copyright law was amended to require royalty payments to composers when their works were mechanically reproduced, and the American Society of Composers, Authors and Publishers (ASCAP) was formed shortly afterward to collect and distribute them. In a highly publicized conversation in *Etude* magazine between Sousa and phonograph inventor Thomas Edison in 1923, the complete reversal in Sousa's tone and the withdrawal of his former arguments after these matters were resolved are striking. A John Philip Sousa that would have been utterly unrecognizable to readers of his earlier article says: "It must not be forgotten that Edison thru the invention of the talking machine has done more to promote good taste in music than any other agency in the world. I have found this particularly emphasized in my own work. Wherever I go with my band, I find that the phonograph has created a lively sense of musical appreciation. People in isolated communities who have never heard a grand opera company, or a symphony orchestra in their lives, thru talking machines and talking machine records, have been able to familiarize themselves with good music."[6] Ethical matters regarding the copying and distribution of mechanically reproduced music were a new set of issues that did need to be addressed when these technologies were introduced, and the outcomes negotiated by Sousa and his colleagues have since provided musicians with financial compensation for the use of their recorded work for more than a century. But as Sousa's apparent softening toward these technologies shows, resolving technological anxiety has far more to do with aligning values and working toward fair solutions than determining whether or not to bar a seemingly threatening technology from the musical landscape.

In the twenty-first century, we feel less fear regarding player pianos and phonographs, now that these anxieties have been addressed and the creative benefits of the recording industry as a whole seem indispensable to our musical endeavors, but technological anxiety in music has not lessened; it has just shifted as we find ourselves in conversations about new technologies. Concerns have emerged on topics such as artist compensation on streaming platforms and the ethics of music made with generative AI, both of which have a great deal in common with the previous worries of the twentieth century about fair pay and credit for musicians' work. Understanding our history with technological anxiety can be a significant boon as we work to navigate today's technological questions. Anxiety and anger emerge when we suspect that our values are misaligned with the values of the people who hold the power to deploy and control new technologies.

So what must we do? First of all, once we understand that it is not automated tools that we actually fear, we must locate the values that underpin our real fears. What are we truly afraid to lose? What does this anxiety reveal to be the thing we value most? Whose competing values are behind the perceived threat? If the people who are managing the new technology have values that align with ours, and if they are working to address our fear, then innovation and democratization can flourish, and our values are clarified and reinforced. But if the people managing the technology have different values, these fears will not be assuaged, and their ongoing presence will signal the fact that there is still a problem to solve. This lack of confidence and trust was the reason for the backlash against Apple's "Crush!" ad. These types of discussions can be messy and complicated. Nevertheless, we must have conversations about how technology is managed and whose values are being served. Throwing out or banning the technology might seem the simplest option, but it is never the best option. Technologies themselves are the product of creative human innovation, and that very ability to dream up new tools and use them to explore our potential is in itself another human value. As Donna Haraway writes, "The machine is not an *it* to be animated, worshipped, and dominated. The machine is us, our processes, an aspect of our embodiment. We can be responsible for machines; *they* do not dominate or threaten us. We are responsible for boundaries; we are they."[7] The only way through these challenges, then, is straight through the middle of the difficult conversations about our values. But if the past shows us anything about these conversations, it is that they are in fact possible, and that positive resolutions can be reached that protect innovation, access, integrity, and fairness. We, too, need to find our way through to our own solutions.

The bottom line of the story of technology and music is simply this: creative humans keep on creating. They create with old tools, new tools, and tools that are developed through the creative process itself. When we fret about the status of "art made by machines" we are losing sight of the actual artist. If art is a communicative act, it is not the machine that is communicating but the artist, who is making creative use of the machine as a tool in human communication.

· · ·

It has been a long journey, crossing more than a century of time and places as we have traveled through the conceptual history of automation technology in music. But stay with me for one last stop, one last moment in music history, in which we can bear witness to the very cultural shift we have been tracing.

Conclusion / 197

It is a rainy April night at the historic Queen Elizabeth Theatre in Vancouver. The lights in the concert hall go down, and the murmuring audience settles into an anticipatory quiet. An automated voice plays over the speakers, introducing the evening's performers in a heavily processed monotone: "Die Mensch-Maschine: Kraftwerk." Wild applause and cheering floods the hall, and four German men file onto the stage as a synthesized beat plays into the darkened space. The members of Kraftwerk take the stage one after the other, faces stoic, gazes downturned, and arrive at their places behind four sleek podiums. These performers will stand in place for the next two hours, performing a string of electronic hits, all without acknowledging their enthusiastic audience's presence one single time during their set.

It is a scene that has played out countless times in concert halls around the globe for more than five decades. But it is not 1975, and the men onstage are not youthful musicians with fresh haircuts and sharp suit jackets. Fifty years have passed since Ralf Hütter and Florian Schneider sparred with journalist Lester Bangs and made North American jaws drop when their thunderous synths panned across the stage during "Autobahn." Their hair is gray now, and when Ralf Hütter intones "Wir fahr'n, fahr'n, fahr'n auf der Autobahn," there is a gravely unsteadiness to his voice. Nobody is bothered by this in the slightest, however. Hütter is seventy-eight years old and is the last remaining member of Kraftwerk's original creative partnership after Florian Schneider's death in 2020. Around me in the concert hall, before and during the show, fans speak almost reverently of the band's legacy and their immeasurable musical influence. More than two thousand people have gathered on this evening in 2025 to revel in that contribution to the music world, celebrating the man-machine collaboration of a now-legendary band. For a group that sparked so much anxiety about the end of human musicianship, it is remarkable to reflect on just how much of it this band has inspired across five decades. Kraftwerk have aged, yes, but the human musicianship behind the machines has endured, and in the popularity of this influential music, we can clearly see what lasts: innovative artistry. A good story. A compelling take on a contemporary question.

In 2025, the synthesizers played by these elderly men don't seem threatening; they are "vintage," and the lyrics about space labs and high-speed trains don't convey futuristic apprehension but nostalgia. As Kraftwerk performs on this night, the visuals projected behind their podiums include simple animations of twentieth-century technologies and black-and-white footage of fashion models and cyclists. Even the robots that appear during the song of the same name are not twenty-first-century robots with lifelike features and smooth movements—they are the plastic-faced, jerky-limbed robots of the 1970s, with their red dress shirts and vacant stares. Decades later, they inspire not apprehension but sentimental recognition and perhaps even a bit of affection for their erratic comportment. Kraftwerk are no longer augurs of a frightening technological future but an anchor to our creative past. Ralf Hütter has stood at his podium, playing live electronics and intoning lyrics into headset microphones on stages around the world for half a century, but the meaning of this music has steadily shifted around him. Kraftwerk, once the curators of unsettling new sounds and modern themes, are now the stewards of an inspirational music legend—and that is a deeply human legacy to inhabit.

At the end of the night, the four performers leave the stage one by one while "Musique Non Stop" plays into the hall. Each one acknowledges the audience for the first time that night during his exit, pausing at the side of the stage to take a slow bow before stepping off. Hütter, appropriately, is the last to leave. Alone on the stage, he plays a final solo before addressing the rapt audience over the steadily looping synthesizers: "Good night," he says softly, "auf wiedersehen." The aging musician leaves his podium and walks to the spotlight, where he pauses to look out over the audience, head high. Cheers and applause erupt from the crowd, and Hütter takes a slow, shallow bow as the audience rise to their feet, shouting and whistling. Half a century after their first international tour, we acknowledge Kraftwerk's story for what it is: their technologically themed music is not a warning about a dehumanized future. It is not a lament for a loss of authentic creativity. It is a communicative act, a deeply human use of technological tools, and a torch passed from one generation of innovators to the next. Hand over his heart, Hütter bows one last time, straightens up, and slowly exits the stage, leaving more than two thousand music lovers to continue their expression of gratitude and awe while Kraftwerk's music loops from the now-empty stage.

Notes

INTRODUCTION

1. "Astonishment as Hologram and a Live Orchestra Put Callas Back Onstage," posted November 29, 2018, by AFP News Agency, YouTube, youtu.be/ieTsKYg1_Qo?
2. Erich Steinhard, "Donaueschingen: Mechanisches Musikfest," *Der Auftakt* 6 (1926): 183, quoted in Thomas Patteson, *Instruments for New Music: Sound, Technology, and Modernism* (University of California Press, 2016), 3–4.
3. Q, "Q Review: Enter Hatsune Miku's Hologram Concert," *CBC Radio*, May 24, 2016, cbc.ca/radio/q/schedule-for-tuesday-may-24-2016-1.3597178/q-review-enter-hatsune-miku-s-hologram-concert-1.3597187.
4. Steinhard, "Donaueschingen," 183.
5. Jessica Riskin, "The Defecating Duck, or, the Ambiguous Origins of Artificial Life," *Critical Inquiry* 29 (Summer 2003): 599–600.
6. Riskin, "Defecating Duck," 609.
7. Jean-Jacques Rousseau, "Essay on the Origin of Language," in *On the Origin of Language*, trans. John H. Moran and Alexander Gode (University of Chicago Press, 1966), 62.
8. Deirdre Loughridge's excellent book, *Sounding Human: Music and Machines, 1740/2020* (University of Chicago Press, 2024), confronts a similar binary, tackling assumptions that underpin the familiar "human *or* machine" logic and working to dismantle narratives of inevitable progression toward machine substitution.
9. Emily I. Dolan, *The Orchestral Revolution: Haydn and the Technologies of Timbre* (Cambridge University Press, 2013), 19–22.
10. John Philip Sousa, *New York Morning Telegraph*, June 12, 1906, quoted in Neil Harris, *Cultural Excursions: Marketing Appetites and Cultural Tastes in Modern America* (University of Chicago Press, 1990), 224.

11. Emily Thompson, "Machines, Music, and the Quest for Fidelity: Marketing the Edison Phonograph in America, 1877–1925," *Musical Quarterly* 79 (1995): 159–60.

12. *National Commission on Technology, Automation, and Economic Progress: Hearings Before the Select Subcommittee on Labor, of the Committee on Education and Labor, House of Representatives, Eighty-eighth Congress, Second Session, on H.R. 10310, and Related Bills to Establish a National Commission on Automation and Technological Progress* (US Government Printing Office, 1964), 96–97.

13. Rick Altman, "Crisis Historiography," in *Silent Film Sound* (Columbia University Press, 2004), 15–26.

14. Lisa Gitelman and Geoffrey Pingree, "What's New About New Media?," in *New Media, 1740–1915* (MIT Press, 2003), xi–xxi.

15. Roland Marchand, *Advertising the American Dream: Making Way for Modernity, 1920–1940* (University of California Press, 1986).

16. Charles McGovern, *Sold American: Consumption and Citizenship, 1890–1945* (University of North Carolina Press, 2006).

CHAPTER 1. THE PLAYER PIANO

Material from this chapter previously appeared in "'This Will Play Your Piano': Automation, Amateur Musicianship, and the Player Piano," *Keyboard Perspectives* 11 (2018): 121–39.

1. Kim Renfro, "Here's Why Modern Songs Play on the Saloon's Piano in 'Westworld,'" *Insider*, October 11, 2016, thisisinsider.com/westworld-piano-songs-2016-10.

2. David Suisman, "Sound, Knowledge, and the 'Immanence of Human Failure': Rethinking Musical Mechanization Through the Phonograph, the Player-Piano, and the Piano," *Social Text* 28 (Spring 2010): 13–14.

3. Works that have examined Stravinsky's relationship with the player piano include Charles M. Joseph, *Stravinsky's Ballets* (Yale University Press, 2011) and Richard Taruskin, *Stravinsky and the Russian Traditions: A Biography of the Works Through* Mavra (University of California Press, 1996). Kyle Gann, *The Music of Conlon Nancarrow* (Cambridge University Press, 1995), examines Nancarrow's works for player piano. Other scholarship that touches on the player piano's amateur users includes Cecilia Björkén-Nyberg, *The Player Piano and the Edwardian Novel* (Ashgate, 2015); Mark Katz, "The Amateur in the Age of Mechanical Music," in *The Oxford Handbook of Sound Studies*, ed. Trevor Pinch and Karin Bijsterveld (Oxford University Press, 2012); Timothy Taylor, "The Commodification of Music at the Dawn of the Era of 'Mechanical Music,'" *Ethnomusicology* 51 (Spring/Summer 2007): 281–305; and Allison Rebecca Wente, *The Player Piano and Musical Labor* (Routledge, 2022).

4. Arthur Loesser, *Men, Women, and Pianos* (Simon and Schuster, 1954), 428.

5. Cyril Ehrlich, *The Piano: A History* (Oxford University Press, 1990), 134.

6. J. P. McEvoy, "The Player-Piano Upstairs," in *Slams of Life, with Malice for All and Charity Toward None* (P. F. Volland, 1919), 22.

7. Harvey Roehl, *Player Piano Treasury*, 2nd ed. (Vestal Press, 1973), 12.

8. Hardman, Peck & Co., advertisement, *New York Times*, October 11, 1925, RPA6.

9. Ehrlich, *Piano*, 136.

10. Sondra Wieland Howe, "Music in the Private Sphere, Churches, and Community," in *Women Music Educators in the United States: A History* (Scarecrow Press, 2014), 51.

11. Mary Burgan, "Heroines at the Piano: Women and Music in Nineteenth-Century Fiction," in *The Lost Chord: Essays on Victorian Music*, ed. Nicholas Temperley (Indiana University Press, 1989), 42.

12. Howe, *Women Music Educators*, 51.

13. Ross Thomson, *Mechanized Shoe ProductionProduction in the United States: The Rise of the Modern Manufacturing System* (Cambridge University Press, 1989), 102; Howe, *Women Music Educators*, 51.

14. Leslie C. Gay Jr., "Before the Deluge: The Technoculture of Song-Sheet Publishing Viewed from Late-Nineteenth-Century Galveston," in *Music and Technoculture*, ed. Rene T. A. Lysloff and Leslie C. Gay Jr. (Wesleyan University Press, 2003), 211.

15. John Greenall, advertisement, *Journal of Domestic Appliances and Sewing Machine Gazette*, January 1, 1886, 33.

16. "Hire Agreements—Important.," *Journal of Domestic Appliances and Sewing Machine Gazette*, July 1, 1886, 13.

17. Petra Meyer-Frazier, "Music, Novels, and Women: Nineteenth-Century Prescriptions for an Ideal Life," *Women and Music: A Journal of Gender and Culture* 10 (2006): 45–46; Charlotte N. Eyerman, "Piano Playing in Nineteenth-Century French Visual Culture," in *Piano Roles: A New History of the Piano*, ed. James Parakilas (Yale University Press, 2001), 180.

18. Harry Braverman, *Labor and Monopoly Capital: The Degradation of Work in the Twentieth Century* (Monthly Review Press, 1974), 188–89.

19. Ernest Newman, *The Piano-Player and Its Music* (Riverside Press, 1920), 18.

20. Newman, *Piano-Player*, 18–19.

21. "First International Sewing Machine and Domestic Appliances Exhibition," *Journal of Domestic Appliances and Sewing Machine Gazette*, December 1, 1887, 24–25.

22. "First International Sewing Machine."

23. Aeolian was founded in 1887; before it became known as a producer of the Pianola, the company built automatic organs. The choice of name likely refers to the Aeolian harp: an instrument that produces sound when air currents pass through it as the wind blows.

24. Wilcox & White Co., advertisement, *New York Times*, May 1, 1898, 20.

25. Aeolian Company, advertisement, *New York Times*, November 23, 1898, 2.

26. Robert Braine, "Self-Playing Pianos," *The Etude* 17, no. 2 (February 1899): 42.

27. Braine, "Self-Playing Pianos."

28. Braine, "Self-Playing Pianos."

29. Annette Richards, "Automatic Genius: Mozart and the Mechanical Sublime," *Music & Letters* 80, no. 3 (August 1999): 366–89; Peter Pesic, "Music of the Clocks and Spheres: Mozart and Haydn's Experiments with Time," *Eighteenth-Century Music* 21 no. 2 (September 2024): 157–85.

30. Carolyn Abbate, "Outside Ravel's Tomb," *Journal of the American Musicological Society* 52 (1999): 476.

31. Emily I. Dolan and John Tresch, "A Sublime Invasion: Meyerbeer, Balzac, and the Opera Machine," *Opera Quarterly* 27, no. 1 (2011): 4–31.

32. Abbate, "Outside Ravel's Tomb," 497.

33. "The Piano Player Vogue," *Musical Opinion & Music Trade Review* 26, no. 304 (January 1903): 309.

34. John Philip Sousa, "The Menace of Mechanical Music," *Appleton's Magazine* 8 (August 1906): 278.

35. Sousa, "Menace of Mechanical Music," 281.

36. 1 Pet. 5:8 (KJV).

37. "The Triumph Piano Player," *Musical Opinion & Music Trade Review*, 27, no. 318 (March 1904): 483.

38. Anti-Pianola, "Mechanical Pianoforte Players," *Musical Opinion & Music Trade Review* 25, no. 290 (November 1901): 101.

39. T. G. Dyson, "Mechanical Pianoforte Players," *Musical Opinion & Music Trade Review* 25 no. 291 (December 1901): 231.

40. William H. Cummings, "Mechanical Music," *Musical Times* (February 1 1905): 93.

41. Cummings, "Mechanical Music," 94.

42. Newman, *Piano-Player*, 11.

43. Newman, *Piano-Player*, 19.

44. Kranich & Bach, advertisement, *American Homes and Gardens*, September 1912, back cover.

45. Lisa Gitelman, "Media, Materiality, and the Measure of the Digital; or, The Case of Sheet Music and the Problem of Piano Rolls," in *Memory Bytes: History, Technology, and Digital Culture*, ed. Lauren Rabinovitz and Abraham Geil (Duke University Press, 2004), 199–217.

46. "Player-Piano Nomenclature: Words to Use and Avoid," *Player Piano* 1, no. 5 (September 1911): 5–6.

47. Advertisers would later find success in deploying similar appeals to taste and control to sell high-fidelity sound equipment in the 1950s. Keir Keightley, "'Turn It Down!' She Shrieked: Gender, Domestic Space, and High Fidelity, 1948–59," *Popular Music* 15 (1996): 149–77. Promotional materials for hi-fi stereo systems encouraged potential customers to think of themselves as being in control of powerful equipment through which they could enter

into transportative musical experiences—not unlike the piano-player, with its promises of personal musical fulfillment.

48. Sydney Grew, *The Art of the Player-Piano: A Textbook for Student and Teacher* (Kegan Paul, Trench, Trubner & Co., 1922), v.
49. Grew, *Art of the Player-Piano*, v–1.
50. Grew, *Art of the Player-Piano*, v–vi.
51. Grew, *Art of the Player-Piano*, 1.
52. Grew, *Art of the Player-Piano*, 7.
53. Grew, *Art of the Player-Piano*, 23.
54. William Delasaire, "Player-Piano Notes," *Musical Times* 66 (July 1, 1925): 620–21.
55. William Delasaire, "Player-Piano Notes," *Musical Times* 66 (October 1, 1925): 918–19.
56. Bertram Smith, "The Piano-Player," *Musical Times* 52 (May 1911): 308–9.
57. Smith, "Piano-Player,", 309.
58. Smith, "Piano-Player,", 309.
59. Smith, "Piano-Player,", 309.
60. Rex Lawson's playing is best witnessed in person, but readers interested in hearing a recording of him performing the second Scherzo at another concert in 2009 can listen at rexlawson.org/music/scherzoBbminor.mp3.

CHAPTER 2. THE REPRODUCING PIANO

1. "Symphony Players in Novel Concert," *New York Sun*, November 18, 1917, 9.
2. "A Notable Presentation of a Notable Instrument," *New York Tribune*, November 25, 1917, 6.
3. Paul Duguid, "Material Matters: The Past and Futurology of the Book," in *The Future of the Book*, ed. Geoffrey Nunberg (University of California Press, 1996), 65.
4. Cyril Ehrlich, *The Piano: A History* (Oxford University Press, 1990), 136.
5. Harvey Roehl, *Player Piano Treasury: The Scrapbook History of the Mechanical Piano in America, As Told in Story, Pictures, Trade Journal Articles and Advertising* (Vestal Press, 1961), 47.
6. Landay Bros., advertisement, *New York Times*, July 27, 1908, 2.
7. Landay Bros., advertisement, *New York Times*, December 24, 1908, 2.
8. Larry Givens, *Re-Enacting the Artist: The Story of the Ampico Reproducing Piano* (Vestal Press, 1970), 90–95.
9. Kent Holliday, "Some American Firms and Their Contributions to the Development of the Reproducing Piano," in *Perspectives on American Music, 1900–1950*, ed. Michael Saffle (Routledge, 2012), 106.
10. Authors are mainly divided between 1905 and 1904. Kent A. Holliday, *Reproducing Pianos Past and Present* (Edwin Mellen Press, 1989), 7;

Arthur W. J. G. Ord-Hume, *Pianola: The History of the Self-Playing Piano* (Allen & Unwin, 1984), 175. "About 1906" is something of an outlier suggested in earlier work. Givens, *Re-Enacting the Artist*, 8–9.

11. Rex Lawson, "On the Right Track: The Recording of Dynamics for the Reproducing Piano (Part One)," *Pianola Journal* 20 (2009): 6.

12. Lawson, "On the Right Track (Part One)," 7.

13. Paul Michael Covey, "Selling 'The Things Money Can't Buy': Piano Advertising in the Mid-Twentieth Century," *Journal of the Society for American Music* 13 (2019): 54–77.

14. Steinway & Sons, advertisement, *Country Life* 31, no. 803 (May 25, 1912): lxxxix.

15. Knabe, Advertisement, *New York Times,* January 15, 1920, 16.

16. Conroy's, Advertisement, *St. Louis Post–Dispatch*, December 11, 1921, B17.

17. Knabe, advertisement, 16; Conroy's, advertisement, B17.

18. United States Department of Labor, *War and Postwar Wages, Prices, and Hours* (US Government Printing Office, 1945), 6.

19. Conroy's, advertisement, B17.

20. Conroy's, advertisement, B17.

21. Knabe, advertisement, *New York Tribune*, March 24, 1918, F10.

22. Bertram Smith, "The Piano-Player," *Musical Times* 52 (May 1, 1911): 309.

23. "Notable Presentation of Notable Instrument," 6.

24. The Welte-Mignon Autograph Piano Co., advertisement, *New York Times*, December 7, 1911, 13.

25. Knabe, advertisement, F10.

26. John Durham Peters, *Speaking into the Air: A History of the Idea of Communication* (University of Chicago Press, 1999), 166.

27. Peters, *Speaking into the Air*, 143.

28. Geo. J. Birkel Co., advertisement, *Los Angeles Times*, November 21, 1909, II.

29. Holliday, *Reproducing Pianos*, 54.

30. Barker Bros., advertisement, *Los Angeles Times*, March 14, 1926, C20.

31. Richard C. Simonton and Ben Hall, liner notes to *The Welte Legacy of Piano Treasures*, Hathaway and Bowers, HSTR-5370, 1963, LP.

32. Holliday, *Reproducing Pianos*, 55–56.

33. Lawson, "On the Right Track (Part One)," 37.

34. Lawson, "On the Right Track (Part One)," 37.

35. Givens, *Re-Enacting the Artist*, 13.

36. Givens, *Re-Enacting the Artist*, 14; Rex Lawson, "On The Right Track—Dynamic Recording for the Reproducing Piano (Part Four)," *Pianola Journal* 23 (2013): 11.

37. Lawson, "On the Right Track (Part Four)," 17.

38. Lawson, "On the Right Track (Part Four)," 17.

39. Givens, *Re-Enacting the Artist*, 28.

40. Lawson, "On the Right Track (Part Four)," 19.
41. Lawson, "On the Right Track (Part Four)," 20.
42. Lawson, "On the Right Track (Part Four)," 20.
43. Holliday, *Reproducing Pianos*, 102.
44. Holliday, *Reproducing Pianos*, 102.
45. Lawson, "On the Right Track (Part Four)," 21.
46. Lawson, "On the Right Track (Part Four)," 21.
47. Lawson, "On the Right Track (Part Four)," 20.
48. The Ampico Corporation, *The AMPICO Service Manual 1929* (American Piano Company, 1929).
49. Givens, *Re-Enacting the Artist*, 25.
50. C. N. Hickman, "A Spark Chronograph Developed for Measuring Loudness of Piano Tones," *Journal of the Acoustical Society of America* 1, no. 35 (1929): 138–46.
51. Hickman, "Spark Chronograph," 141–42.
52. Hickman, "Spark Chronograph," 142.
53. American Piano Company, advertisement, *Tuners Journal* 8, no. 13 (June 1929).
54. American Piano Company, advertisement.
55. Nick Seaver, "'This Is Not a Copy': Mechanical Fidelity and the Re-enacting Piano," *Differences* 22 (2011): 60–61.
56. Seaver, "'This Is Not a Copy,'" 64–65.
57. Seaver, "'This Is Not a Copy,'" 56.
58. Rex Lawson, "On the Right Track (Part Four)," 25–26.
59. Holliday, *Reproducing Pianos*, 54.
60. The Aeolian Company, advertisement, *New York Times*, May 23, 1917, 7.
61. The Aeolian Company, advertisement, *Cosmopolitan* 61, no. 5 (October 1916): 92–93.

CHAPTER 3. SYNTHESIZERS

1. Pascal Bussy, *Kraftwerk: Man, Machine, and Music* (SAF Publishing, 1993), 138.
2. Chris Morris, "Wendy Carlos Takes Her Moog Music to East Side Digital," *Billboard*, October 3, 1998, 69.
3. For a general overview of the history of German electronic music and Kraftwerk's influence, see Sean Nye, "Minimal Understandings: The Berlin Decade, the Minimal Continuum, and Debates on the Legacy of German Techno," *Journal of Popular Music Studies* 25 (June 2013): 154–84. Mark Duffett, "Average White Band: Kraftwerk and the Politics of Race," in *Kraftwerk: Music Non-Stop*, ed. Sean Albiez and David Pattie (Continuum, 2011), 194–213, looks back to the group's influences, from The Velvet Underground to James Brown. Michael K. Bourdaghs, "Happy End, Arai Yumi, and Yellow Magic Orchestra," in *Sayonara Amerika, Sayonara Nippon: A Geopolitical

Prehistory of J-pop (Columbia University Press, 2012), 159–94, situates Yellow Magic Orchestra as a part of a diverse scene of popular "new music" in Japan in the 1970s, which would pave the way for much contemporary Japanese pop.

4. Alvin Toffler, *Future Shock* (Random House, 1970).

5. Douglas Rushkoff, *Present Shock: When Everything Happens Now* (Penguin Group, 2013), 14.

6. Jon Savage, "Interview: Kraftwerk," Red Bull Music Academy, August 30, 2012, daily.redbullmusicacademy.com/2012/08/kraftwerk-interview, accessed May 28, 2019.

7. Rob Johnstone, *Kraftwerk and the Electronic Revolution* (2008; Films Media Group, 2013), eVideo.

8. Wolfgang Flür et al., *Kraftwerk: I Was a Robot* (Omnibus Press, 2017), 76.

9. Flür et al., *Kraftwerk*.

10. Flür et al., *Kraftwerk*, 80–90.

11. Mitchell Schneider, "The Man-Machine," *Rolling Stone*, May 18, 1978, rollingstone.com/music/music-album-reviews/the-man-machine-96960. Given that many of the synths Kraftwerk used, such as the Minimoog, are perceived now as having a "warm" sound, this initial "cold" reception provides an example of changing perceptions of musical technologies.

12. Theo Cateforis, *Are We Not New Wave? Modern Pop at the Turn of the 1980s* (University of Michigan Press, 2011), 151.

13. John Doran, "Karl Bartos Interviewed: Kraftwerk and the Birth of the Modern," *The Quietus*, March 11, 2009, thequietus.com/articles/01282-karl-bartos-interviewed-kraftwerk-and-the-birth-of-the-modern, accessed May 27, 2019.

14. Sean Albiez and Kyrre Tromm Lindvig, "*Autobahn* and Heimatklänge: Soundtracking the FRG," in *Kraftwerk: Music Nonstop*, ed. Sean Albiez and David Pattie (Continuum, 2011), 15.

15. Ulrich Adelt, *Krautrock: German Music in the Seventies* (University of Michigan Press, 2016), 27.

16. Lester Bangs, "Kraftwerk: The Final Solution to the Music Problem?," *New Musical Express*, September 6, 1975, 20–21.

17. "The Piano Player Vogue," *Musical Opinion & Music Trade Review* 26, no. 304 (January 1903): 309.

18. Bangs, "Kraftwerk," 20.

19. Ralf Hütter, "Interview," WSKU (Kent–Ohio), first broadcast June 19, 1978, cited in David Pattie, "Introduction: The (Ger)Man Machines," in *Kraftwerk: Music Non-stop*, ed. Sean Albiez and David Pattie (Continuum: 2011), 2.

20. David Pattie, "Introduction: The (Ger)Man Machines," in *Kraftwerk: Music Non-stop*, ed. Sean Albiez and David Pattie (Continuum, 2011), 9.

21. Jose Ortega y Gasset, *El Espectador III*, 12, quoted in M. J. Neves, "The Dehumanization of Art: Ortega y Gasset's Vision of New Music," *International Review of the Aesthetics and Sociology of Music* 43 (December 2012): 369.

22. Keir Keightley, "Reconsidering Rock," in *The Cambridge Companion to Pop and Rock*, ed. Simon Frith, Will Straw, and John Street (Cambridge University Press, 2001), 109–42.
23. Keightley, "Reconsidering Rock," 135.
24. Bangs, "Kraftwerk," 21.
25. Bangs, "Kraftwerk," 21.
26. Ralf Hütter, "Interview," Beacon Radio, Birmingham, UK, June 14, 1981, archive.org/web/20081202043525/http://kraftwerk.technopop.com.br/interview_122.php .
27. Mike Beecher, "Kraftwerk Revealed: An Interview with Ralf Hütter," *Electronics & Music Maker*, September 1981, 62–63.
28. David Buxton, "Music, the Star System, and the Rise of Consumerism," in *On Record: Rock, Pop and the Written Word*, ed. Simon Frith and Andrew Goodwin (Routledge, 2006), 427–40.
29. Bussy, *Kraftwerk*, 64.
30. Bob Doerschuk, "Orchestral Manoeuvres in the Dark," *Keyboard*, April 1982, 30.
31. Cateforis, *Are We Not New Wave?*, 169.
32. David Pattie, "Kraftwerk: Playing the Machines," in *Kraftwerk: Music Non-Stop*, ed. Sean Albiez and David Pattie (Continuum: 2011), 119.
33. Gronholm notes that Flür and Bartos were hired on the basis of their ability to control their movements in performance and refrain from playing dramatically. Johannes Gronholm, "Kraftwerk," in *Kraftwerk: Music Non-Stop*, ed. Sean Albiez and David Pattie (Continuum, 2011), 77.
34. Flür et al., *Kraftwerk*, 139–40.
35. Flür et al., *Kraftwerk*, 141.
36. Bangs, "Kraftwerk," 30–31.
37. Michael K. Bourdaghs, *Sayonara Amerika, Sayonara Nippon: A Geopolitical Prehistory of J-Pop* (Columbia University Press, 2012), 143.
38. James Henke, "Yellow Magic Orchestra: Japanese Technopop Is Poised to Invade America," *Rolling Stone*, no. 319, June 12, 1980.
39. Minoru Inaba, "Computer Rock Music Gaining Fans," *Sarasota Journal*, August 18, 1980, 11A.
40. Harry Sumrall, "Yellow Magic," *Washington Post*, November 6, 1979, B7.
41. "Yellow Magic Orch," *Variety*, December 17, 1980, 72.
42. Inaba, "Computer Rock Music Gaining Fans," 11A.
43. Like other music-performance programs such as *American Bandstand*, most artists appearing on *Soul Train* mimed to prerecorded tracks, and YMO did the same during this appearance.
44. James Hadfield, "YMO's Yukihiro Takahashi Celebrates a Pair of 40th Anniversaries," *Japan Times*, November 23, 2018, japantimes.co.jp/culture/2018/11/23/music/ymos-yukihiro-takahashi-celebrates-pair-40th-anniversaries/#.XQhczBZKhaQ.

45. Neil Lerner, "Mario's Dynamic Leaps: Musical Innovations (and the Specter of Early Cinema) in *Donkey Kong* and *Super Mario Bros.*," in *Music in Video Games: Studying Play*, ed. K. J. Donnelly et al. (Routledge, 2014), 1.

46. Bourdaghs, *Sayonara Amerika, Sayonara Nippon*, 144; Hugh Kenner, *The Counterfeiters: An Historical Comedy* (Dalkey Archive Press, 2005).

47. Brian Currid, "'Finally, I Reach to Africa': Ryuichi Sakamoto and Sounding Japan(ese)," in *Contemporary Japan and Popular Culture*, ed. John Whittier Treat (University of Hawaii Press, 1996), 75.

48. Currid, "'Finally, I Reach to Africa,'" 145.

49. Mark Butler, *Playing with Something That Runs: Technology, Improvisation, and Composition in DJ and Laptop Performance* (Oxford University Press, 2014), 95–96.

50. Butler, *Playing with Something That Runs*, 101–2.

CHAPTER 4. DRUM MACHINES

1. Dan LeRoy, *Dancing to the Drum Machine* (Bloomsbury Academic, 2023), 143–44.

2. Margaret Schedel, "Anticipating Interactivity: Henry Cowell and the Rhythmicon," *Organised Sound* 7, no. 3 (December 2002): 247–54.

3. Henry Cowell, *New Musical Resources* (Alfred A. Knopf, 1930).

4. Nicolas Slonimsky, "Henry Cowell," in *American Composers on American Music: A Symposium*, ed. Henry Cowell (Frederick Ungar, 1962), 59–60.

5. Wurlitzer, advertisement, *International Musician* 59, no. 8 (February 1961): 6.

6. Wurlitzer, advertisement, *International Musician* 59, no. 8 (February 1961): 6.

7. Robert Michler has provided a summary of a handful of songs from 1969 to 1976 that nevertheless did use this generation of drum machines in recordings, with interesting artistic results. See Robert Michler, "A Romanticized Narrative and the Overlooked Birth of Electronic Beats" in *Studies in the Arts II—Kunste, Design und Wissenschart im Austausch*, ed. Thomas Gartmann et al. (Transcript Vertag, 2023).

8. John Schaefer, "Shuggie Otis Spreads His 'Wings,' 40 Years Later," New Sounds, April 16, 2013, newsounds.org/story/287163-shuggie-otis-wings.

9. Rick Mattingly, "Roger Linn," *Modern Drummer*, February/March 1982, 20, 100.

10. Rick Mattingly, "Roger Linn," 100.

11. Stanley Hall, "Simon Phillips," *Modern Drummer*, June 1981, 11.

12. Robyn Flans, "Harvey Mason," *Modern Drummer*, July 1981, 64.

13. "Editor's Overview," *Modern Drummer*, February/March 1982, 2.

14. Robert Carr, "The Linn LM-1: How It Works, What It Can Do," *Modern Drummer*, February/March 1982, 18, 96–97.

15. Carr, "Linn LM-1," 18.

16. Aeolian Company, Advertisement, *New York Tribune*, October 30, 1898, B3.
17. Robyn Flans, "Jeff Porcaro," *Modern Drummer*, February/March 1982, 19.
18. Robyn Flans, "Jeff Porcaro," *Modern Drummer*, February 1983, 46.
19. Flans, "Jeff Porcaro," February/March 1982, 19. Subsequent quotations are from the same article.
20. Robyn Flans, "Jim Keltner," *Modern Drummer*, February/March 1982, 19–20. Subsequent quotations are from the same article.
21. Flans, "Jim Keltner," 19.
22. Rick Mattingly, "Mel Lewis," *Modern Drummer*, February 1985, 50.
23. Mattingly, "Mel Lewis," 8.
24. Rick Mattingly, "Terry Bozzio," *Modern Drummer*, December 1984: 60.
25. Robyn Flans, "Narada Michael Walden: Inspired," *Modern Drummer*, April 1982, 29.
26. Robyn Flans, "Narada Michael Walden," 86.
27. Robyn Flans, "Narada Michael Walden," 86.
28. Margie Borschke, "Disco Edits and Their Discontents: The Persistence of the Analog in a Digital Era," *New Media & Society* 13, no. 6 (2011): 929–44.
29. Flans, "Jeff Porcaro," February/March 1982, 19.
30. Joe Mansfield, *Beat Box: A Drum Machine Obsession* (Get On Down, 2013), 124.
31. Thomas Brett, "Prince's Rhythm Programming: 1980s Music Production and the Esthetics of the LM-1 Drum Machine," *Popular Music and Society* 43, no. 3 (2020): 244–61.
32. Robyn Flans, "Classic Tracks: Phil Collins' 'In the Air Tonight,'" *Mix*, May 1, 2005, mixonline.com/mag/audio_phil_collins_air/index.html.
33. Andy Greene, "Phil Collins: My Life in 15 Songs," *Rolling Stone*, February 29, 2016, rollingstone.com/music/music-lists/phil-collins-my-life-in-15-songs-82641.
34. Robynn J. Stilwell, "In the Air Tonight: Text, Intertextuality, and the Construction of Meaning," *Popular Music and Society* 19, no. 4 (1995): 93–97.
35. "Roger Linn: Technical GRAMMY Award Acceptance," *Grammy Awards*, February 13, 2011, grammy.com/videos/roger-linn-at-special-merit-awards-ceremony-nominees-reception.
36. Stephen Bidwell, "Close Your Eyes, and Listen: Death Cab for Cutie's Jason McGerr," *Modern Drummer*, January 2019, 48.

CHAPTER 5. SINGING SYNTHESIS

1. Thomas L. Hankins and Robert J. Silverman, *Instruments and Imagination* (Princeton University Press, 2014), 193.
2. Wolfgang von Kempelen, *Mechanismus der menschlichen Sprache nebst der Beschreibung einer sprechenden Machine* (J. B. Degen, 1791), cited in Willem Levelt, *A History of Psycholinguistics: The Pre-Chomskyan Era* (Oxford University Press, 2013), 129–31.

3. "Kempelen's Speaking Machine," posted June 6, 2017, by Fabian Brackhane, YouTube, youtube.com/watch?v=k_YUB_S6Gpo.

4. Dennis H. Klatt, "Review of Text-to-Speech Conversion for English," *Journal of the Acoustical Society of America* 82 (1987): 737–93.

5. Readers can enjoy the IBM 7094's performance at youtu.be/41U78QP8nBk.

6. Scott Wilkinson, "Humanoid or Vocaloid?," *Electronic Musician*, August 1, 2003, emusician.com/mag/emusic_humanoid_vocaloid/index.html.

7. Keisuke Yamada, *Supercell Featuring Hatsune Miku* (Bloomsbury Academic, 2017), 16.

8. Peter Manning, *Electronic and Computer Music* (Oxford University Press, 2013), 471–82.

9. Yamada, *Supercell Featuring Hatsune Miku*, 13–14.

10. Manning, *Electronic and Computer Music*, 82.

11. Yamada, *Supercell Featuring Hatsune Miku*, 14.

12. "ボカロ 2013 年 月別投稿者数グラフ" (Graph of 2013 monthly Vocaloid uploader counts), *NicoNico*, posted January 4, 2014, ch.nicovideo.jp/kadotanimitsuru/blomaga/ar428894, .

13. Lucy Bennett, "Tracing Textual Poachers: Reflections on the Development of Fan Studies and Digital Fandom," *Journal of Fandom Studies* 2, no. 1 (2014): 5–20.

14. Alvin Toffler, *The Third Wave* (Bantam Books, 1980), 265–75.

15. The animation was a parody of a popular meme derived from the anime *Bleach*, in which a character was depicted spinning a green onion while the Finnish song "Ievan Polkka" played in the background.

16. Trevor J. Pinch and Wiebe E. Bijker, "The Social Construction of Facts and Artefacts: Or How the Sociology of Science and the Sociology of Technology Might Benefit Each Other," *Social Studies of Science* 14 (August 1984): 399-441.

17. Trevor Pinch and Frank Trocco, *Analog Days: The Invention and Impact of the Moog Synthesizer* (Harvard University Press, 2002).

18. Song titles in Japanese are given in their original script when applicable, followed by a romanized transliteration and an English translation upon first mention. Thereafter, I use the version of the title most commonly used by the English-language Vocaloid community.

19. Masataka Goto, "OngaCREST Project: Building a Similarity-Aware Information Environment for a Content-Symbiotic Society," in *Human-Harmonized Information Technology*, vol. 2, *Horizontal Expansion*, ed. Toyoaki Nishida (Springer Japan, 2017), 1–2.

20. Goto, "OngaCREST Project," 3.

21. Songrium, accessed April 26, 2025, songrium.jp.

22. Songrium, accessed April 26, 2025, songrium.jp.

23. Henry Jenkins, *Confronting the Challenges of Participatory Culture: Media Education for the 21st Century* (MIT Press, 2009), 7.

24. Aaron Delwiche and Jennifer Jacobs Henderson, "Introduction: What Is Participatory Culture?," in *The Participatory Cultures Handbook*, ed. Jennifer Jacobs Henderson and Aaron Alan Delwiche (Routledge, 2013), 7.

25. Delwiche and Jennifer Jacobs Henderson, "Introduction," 7.

26. Sherry Turkle, *Alone Together: Why We Expect More from Technology and Less from Each Other* (Basic Books, 2011).

27. See Alison Slade et al., eds., *Television, Social Media, and Fan Culture* (Lexington Books, 2015) and Kristin M. Barton and Jonathan Malcolm Lampley, eds., *Fan CULTure: Essays on Participatory Fandom in the 21st Century* (McFarland, 2014).

28. Supercell, "Discography," accessed July 8, 2019, supercell.jp/discography.html.

29. Oricon News, "Supercell のメンバーにインタビュー" [Interview with Supercell members], accessed December 11, 2018, oricon.co.jp/trend/hayari/20080805_02.html.

30. "初音ミク が オリジナル曲を歌ってくれたよ「メルト」,("Hatsune Miku sang original song'Melt'," posted December 7, 2007, by Ryo, *NicoNico*, nicovideo.jp/watch/sm1715919. "Miku Miku Ni Shite Ageru" continues to hold the number one ranking as of 2025.

31. Yamada, *Supercell Featuring Hatsune Miku*, 51.

32. Supercell, "Biography," accessed July 8, 2019, supercell.jp/biography.html.

33. Masataka Goto, "Frontiers of Music Technologies: Singing Synthesis and Active Music Listening" (presentation at the CIRMMT Distinguished Lecture, Montreal, QC, September 24, 2015).

CHAPTER 6. HOLOGRAPHIC PERFORMANCE

Material from this chapter appeared previously in "Locating Liveness in Holographic Performances: Technological Anxiety and Participatory Fandom at Vocaloid Concerts," *Popular Music* 41, no. 1 (2022): 1–19. Reprinted with permission.

1. Some fans use chemical glow sticks at these concerts, but battery-powered LED light sticks are increasingly common. In Japan, the light sticks are called penlights (ペンライト), and some fans use this term in English as well.

2. Leon Weinstein, "Hatsune Miku at the Sony Centre for the Performing Arts," *Live in Limbo*, May 23, 2016, liveinlimbo.com/2016/05/23/concert-reviews/hatsune-at-the-sony-centre-for-the-performing-arts.html.

3. Cory Garcia, "Hatsune Miku's Hologram Concert Is a Sure Sign of Endtimes, and I Can't Wait," *Dallas Observer*, November 17, 2015, dallasobserver.com/music/hatsune-mikus-hologram-concert-is-a-sure-sign-of-endtimes-and-i-cant-wait-7780679.

4. Keisuke Yamada, *Supercell Featuring Hatsune Miku* (Bloomsbury Academic, 2017); Masataka Goto, "OngaCREST Project: Building a Similarity-Aware Information Environment for a Content-Symbiotic Society," in

Human-Harmonized Information Technology, vol. 2, *Horizontal Expansion*, ed. Toyoaki Nishida (Springer Japan, 2017); Nina Sun Eidsheim, "Race as Zeroes and Ones: Vocaloid Refused, Reimagined, and Repurposed," in *The Race of Sound: Listening, Timbre, and Vocality in African-American Music* (Duke University Press, 2019), 115–50; Yiyi Yin, "Vocaloid in China: Cosmopolitan Music, Cultural Expression, and Multilayer Identity," *Global Media and China* 3, no. 1 (2018): 51–66.

5. Stina Hasse Jørgensen et al., "Hatsune Miku: An Uncertain Image," *Digital Creativity* 28, no. 4 (2017): 318–31.

6. Yasuhiro Arai and Shinya Kinukawa, "Copyright Infringement as User Innovation," *Journal of Cultural Economics* 38, no. 2 (2014): 131–44.

7. For more on participatory fandom, see Kristin Michael Barton and Jonathan Malcolm Lampley, eds., *Fan CULTure: Essays on Participatory Fandom in the 21st Century* (McFarland, 2014) and Henry Jenkins, *Fans, Bloggers, and Gamers: Exploring Participatory Culture* (New York University Press, 2006).

8. Rafal Zaborowski, "Hatsune Miku and Japanese Virtual Idols," in *The Oxford Handbook of Music and Virtuality*, ed. Sheila Whiteley and Shara Rambarran (Oxford University Press, 2016), 121.

9. Zaborowski, "Hatsune Miku and Japanese Virtual Idols," 124.

10. Hatsune Miku began performing onstage at events in Japan and Singapore in 2009. Her first appearance in the West, at a 2011 show in Los Angeles, preceded the Miku Expo touring events by three years.

11. Calum Marsh, "We Attended the Hatsune Miku Expo to Find Out If a Hologram Pop Star Could Be Human," *Vice*, May 27, 2016, vice.com/en_us/article/ae87yb/hatsune-miku-expo-feature.

12. Mary Lucus-Flannery, "What I Learned from a Vocaloid and 1500 Screaming Millennials," *Medium*, May 5, 2016, medium.com/@marylucusflannery/what-i-learned-from-a-vocaloid-and-1500-screaming-millennials-d181a26d5ce6.

13. Lucy Bennett, "Tracing Textual Poachers: Reflections on the Development of Fan Studies and Digital Fandom," *Journal of Fandom Studies* 2, no. 1 (2014): 5–20.

14. Matthew Reason and Anja Mølle Lindelof, *Experiencing Liveness in Contemporary Performance: Interdisciplinary Perspectives* (Routledge, 2014).

15. Daniel Cavicchi, *Tramps Like Us: Music and Meaning Among Springsteen Fans* (Oxford University Press, 1998).

16. Reason and Lindelof, *Experiencing* , 12.

17. In 2022, ABBA rejected this label when journalists applied it to their new avatar concert, ABBA Voyage, in an effort to distinguish between the transparent-screen projections and their own LED screen.

18. Christine R. Yano, "Letters from the Heart: Negotiating Fan-Star Relationships in Japanese Popular Music," in *Fanning the Flames: Fans and Consumer Culture in Contemporary Japan*, ed. William W. Kelly (SUNY Press, 2004), 41–58; Hiroshi Aoyagi, "Pop Idols and the Asian Identity," in *Japan Pop!*

Inside the World of Japanese Pop Culture, ed. Timothy J. Craig (M. E. Sharpe, 2000); Hiroshi Aoyagi, *Islands of Eight Million Smiles: Idol Performance and Symbolic Production in Contemporary Japan* (Harvard University Press, 2005).

19. Bruce A. Austin, "Portrait of a Cult Film Audience: *The Rocky Horror Picture Show*," Journal of Communication 31, no. 2 (1981): 43–54.

20. Clifford Geertz, "Religion as a Cultural System," in *The Interpretation of Cultures* (Basic Books, 1973), 87–125.

21. Patrick Macias, "CRN Interview: The Creators of Hatsune Miku," *Crunchyroll*, July 21, 2011, crunchyroll.com/anime-feature/2011/07/21/crn-interview-the-creators-of-hatsune-miku; NBT and Sizergyia, "Interview with the Creator of Hatsune Miku, Hiroyuki Itoh," *JRock News*, March 11, 2019, jrocknews.com/2019/03/interview-hiroyuki-itoh-creator-of-hatsune-miku.html.

22. NBT and Sizergyia, "Interview with the Creator of Hatsune Miku."

23. Paul Sanden, "Virtual Liveness and Sounding Cyborgs: John Oswald's 'Vane,'" *Popular Music* 31, no. 1 (2012): 45–68.

24. Sanden, "Virtual Liveness and Sounding Cyborgs," 50.

25. Eduardo Coutinho and Klaus R. Scherer, "The Effect of Context and Audio-Visual Modality on Emotions Elicited by a Musical Performance," *Psychology of Music* 45, no. 4 (2017): 550–69, shows the difference in feelings of awe and empathy at live performances versus prerecorded

26. Brandon Wetherbee, "Live DC: Hatsune Miku Expo @ The Anthem," *Brightest Young Things*, July 13, 2018, brightestyoungthings.com/articles/live-dc-hatsune-miku-expo-the-anthem.

27. Martin Barker, "Coming A(live): A Prolegomenon to any Future Research on 'Liveness,'" in *Experiencing Liveness in Contemporary Performance: Interdisciplinary Perspectives*, ed. Matthew Reason and Anja Mølle Lindelof (Routledge, 2017), 21–33.

28. Alexander Sigman, "Robot Opera: Bridging the Anthropocentric and the Mechanized Eccentric," *Computer Music Journal* 43, no. 1 (2019): 21–37.

29. Dwango Co., "NicoNico Chokaigi," 2019, chokaigi.jp/2019/en/about.

30. While Chokaigi has returned following the COVID-19 pandemic, Cho Party has yet to be staged postpandemic as of 2025.

31. "初音ミクオリジナル曲「初音ミクの消失 (LONG VERSION)」." ("An Original Hatsune Miku Song [The Disappearance of Hatsune Miku (LONG VERSION)])," *NicoNico*, posted by cosMo@暴走P, April 8, 2008, nicovideo.jp/watch/sm2937784.

32. To watch a group of vocal coaches put together a highly impressive performance, see nicovideo.jp/watch/so40371474.

33. For discussion of glitch aesthetics, see Carolyn L. Kane, *High-Tech Trash: Glitch, Noise, and Aesthetic Failure* (University of California Press, 2019).

34. Tito Velasquez, Copenhagen Virtual Concert party organizer, interview by John Glanville, research assistant, May 21, 2023.

CONCLUSION

1. Devin Coldewey, "Apple's 'Crush' Ad Is Disgusting," *TechCrunch*, May 9, 2024, https://techcrunch.com/2024/05/09/apples-crush-ad-is-disgusting/; Michelle Hawley, "Apple's 'Crush' Ad Controversy: Insights into the Power of Public Opinion," *CMSWire*, May 10, 2024, https://www.cmswire.com/digital-marketing/apples-crush-ad-controversy-insights-into-the-power-of-public-opinion/; Leander Kahney, "Apple's 'Crush!' iPad Ad Draws Fire," *Cult of Mac Today*, May 8, 2024, https://newsletters.cultofmac.com/p/apples-crush-ipad-ad-draws-fire.

2. Tripp Mickle, "Apple Says Destructive iPad Ad 'Missed the Mark,'" *The New York Times*, May 9, 2024, https://www.nytimes.com/2024/05/09/technology/apple-ipad-ad-crush-apology.html .

3. "Hydraulic Press Channel," *Social Blade*, accessed April 14, 2025, https://socialblade.com/youtube/channel/UCcMDM0Nu66_1Hwi5-MeiQgw/monthly.

4. Peter C. Baker, "That Much-Despised Apple Ad Could Be More Disturbing Than It Looks," *New York Times Magazine*, June 6, 2024, https://www.nytimes.com/2024/06/06/magazine/apple-ipad-ad.html.

5. Patrick Warfield, "John Philip Sousa and 'The Menace of Mechanical Music,'" *Journal for the Society of American Music* 3, no. 4 (2009): 431–63.

6. "A Momentous Musical Meeting," *Etude* (October 1923): 663.

7. Donna J. Haraway, *Simians, Cyborgs, and Women: The Reinvention of Nature* (Routledge, 1991), 181.

Bibliography

Abbate, Carolyn. "Outside Ravel's Tomb." *Journal of the American Musicological Society* 52, no. 3 (1999): 465–530.
Adelt, Ulrich. *Krautrock: German Music in the Seventies*. University of Michigan Press, 2016.
Aeolian Company. Duo-Art Advertisement. *Cosmopolitan* 61, no. 5 (October 1916): 92–93.
Aeolian Company. Duo-Art Advertisement. *New York Times*, May 23, 1917, 7.
Aeolian Company. Pianola Advertisement. *New York Times*, November 23, 1898, 2.
Aeolian Company. Pianola Advertisement. *New York Tribune*, October 30, 1898, B3.
Aeolian Company. Duo-Art Advertisement. *Atlantic Monthly*, December 1922, 97–98.
Albiez, Sean, and Kyrre Tromm Lindvig. "Autobahn and Heimatklange: Soundtracking the FRG." In *Kraftwerk: Music Non-stop*, edited by Sean Albiez and David Pattie, 13–30. Continuum, 2011.
Albiez, Sean, and David Pattie, eds. *Kraftwerk: Music Non-Stop*. Continuum, 2011.
Altman, Rick. "Crisis Historiography." In *Silent Film Sound*, 15–26. Columbia University Press, 2004.
American Piano Company. Advertisement. *Tuners Journal* 8, no. 13 (June 1929).
The Ampico Corporation. *The AMPICO Service Manual 1929*. American Piano Company, 1929.
Anti-Pianola. "Mechanical Pianoforte Players." *Musical Opinion & Music Trade Review* 25, no. 290 (November 1901): 101.
Aoyagi, Hiroshi. *Islands of Eight Million Smiles: Idol Performance and Symbolic Production in Contemporary Japan*. Harvard University Press, 2005.

Aoyagi, Hiroshi. "Pop Idols and the Asian Identity." In *Japan Pop! Inside the World of Japanese Pop Culture*, edited by Timothy J. Craig, 77–92. M. E. Sharpe, 2000.

Arai, Yasuhiro, and Shinya Kinukawa. "Copyright Infringement as User Innovation." *Journal of Cultural Economics* 38, no. 2 (2014): 131–44.

"Astonishment as Hologram and a Live Orchestra Put Callas Back Onstage." Posted November 29, 2018, by AFP News Agency. YouTube. https://youtu.be/ieTsKYg1_Qo?

Austin, Bruce A. "Portrait of a Cult Film Audience: *The Rocky Horror Picture Show*." *Journal of Communication* 31, no. 2 (1981): 43–54.

Baker, Peter C. "That Much-Despised Apple Ad Could Be More Disturbing Than It Looks." *New York Times Magazine*, June 6, 2024. https://www.nytimes.com/2024/06/06/magazine/apple-ipad-ad.html.

Bangs, Lester. "KRAFTWERKFEATURE, or, How I Learned to Stop Worrying & Love the Bahn." *Creem*, September 1975, 30–31.

Barker Bros. Advertisement. *Los Angeles Times*, March 14, 1926, C20.

Barker, Martin. "Coming A(live): A Prolegomenon to Any Future Research on 'Liveness.'" In *Experiencing Liveness in Contemporary Performance: Interdisciplinary Perspectives*, edited by Matthew Reason and Anja Mølle Lindelof, 21–33. Routledge, 2017.

Barton, Kristin M., and Jonathan Malcolm Lampley, eds. *Fan CULTure: Essays on Participatory Fandom in the 21st Century*. McFarland, 2014.

Beecher, Mike. "Kraftwerk Revealed: An Interview with Ralf Hütter." *Electronics & Music Maker*, September 1981, 62–63.

Bennett, Lucy. "Tracing Textual Poachers: Reflections on the Development of Fan Studies and Digital Fandom." *Journal of Fandom Studies* 2, no. 1 (2014): 5–20.

Bidwell, Stephen. "Close Your Eyes, and Listen: Death Cab for Cutie's Jason McGerr." *Modern Drummer*, January 2019, 48.

Björkén-Nyberg, Cecilia. *The Player Piano and the Edwardian Novel*. Ashgate, 2015.

Bokaro 2013 tsukibetsu toukoushasuu gurafu (ボカロ 2013 年 月別投稿者数グラフ) [Graph of 2013 monthly Vocaloid uploader counts]. *NicoNico*. Posted January 4, 2014. ch.nicovideo.jp/kadotanimitsuru/blomaga/ar428894.

Borschke, Margie. "Disco Edits and Their Discontents: The Persistence of the Analog in a Digital Era." *New Media & Society* 13, no. 6 (2011): 929–44.

Bourdaghs, Michael K. "Happy End, Arai Yumi, and Yellow Magic Orchestra." In *Sayonara Amerika, Sayonara Nippon: A Geopolitical Prehistory of J-pop*, 159–94. Columbia University Press, 2012.

Bourdaghs, Michael K. *Sayonara Amerika, Sayonara Nippon: A Geopolitical Prehistory of J-Pop*. Columbia University Press, 2012.

Braine, Robert. "Self-Playing Pianos." *The Etude* 17, no. 2 (February 1899): 42.

Braverman, Harry. *Labor and Monopoly Capital: The Degradation of Work in the Twentieth Century*. Monthly Review Press, 1974.

Brett, Thomas. "Prince's Rhythm Programming: 1980s Music Production and the Esthetics of the LM-1 Drum Machine." *Popular Music and Society* 43, no. 3 (2020): 244–61.

Burgan, Mary. "Heroines at the Piano: Women and Music in Nineteenth-Century Fiction." In *The Lost Chord: Essays on Victorian Music*, edited by Nicholas Temperley, 39–55. Indiana University Press, 1989.

Bussy, Pascal. *Kraftwerk: Man, Machine, and Music*. SAF Publishing, 1993.

Butler, Mark. *Playing with Something That Runs: Technology, Improvisation, and Composition in DJ and Laptop Performance*. Oxford University Press, 2014.

Buxton, David. "Music, the Star System, and the Rise of Consumerism." In *On Record: Rock, Pop and the Written Word*, edited by Simon Frith and Andrew Goodwin, 427–40. Routledge, 2006.

Carr, Robert. "The Linn LM-1: How It Works, What It Can Do." *Modern Drummer*, February/March 1982, 18, 96–97.

Cateforis, Theo. *Are We Not New Wave? Modern Pop at the Turn of the 1980s*. University of Michigan Press, 2011.

Cavicchi, Daniel. *Tramps Like Us: Music and Meaning Among Springsteen Fans*. Oxford University Press, 1998.

Coldewey, Devin. "Apple's 'Crush' Ad Is Disgusting." *TechCrunch*, May 9, 2024. https://techcrunch.com/2024/05/09/apples-crush-ad-is-disgusting/.

Conroy's. Advertisement. *St. Louis Post-Dispatch*, December 11, 1921, B17.

Coutinho, Eduardo, and Klaus R. Scherer. "The Effect of Context and Audio-Visual Modality on Emotions Elicited by a Musical Performance." *Psychology of Music* 45, no. 4 (2017): 550–69.

Covey, Paul Michael. "Selling 'The Things Money Can't Buy': Piano Advertising in the Mid-Twentieth Century." *Journal of the Society for American Music* 13 (2019): 54–77.

Cowell, Henry. *New Musical Resources*. Alfred A. Knopf, 1930.

Cummings, William H. "Mechanical Music." *Musical Times*, February 1, 1905, 93.

Currid, Brian. "'Finally, I Reach to Africa': Ryuichi Sakamoto and Sounding Japan(ese)." In *Contemporary Japan and Popular Culture*, edited by John Whittier Treat, 74–98. University of Hawaii Press, 1996.

Dancis, Bruce. "Gary Numan: Britain's New Wave Techno-Rocker." *Contemporary Keyboard*, August 1980, 39.

Delasaire, William. "Player-Piano Notes." *Musical Times* 66 (July 1, 1925): 620–21.

Delasaire, William. "Player-Piano Notes." *Musical Times* 66 (October 1, 1925): 918–19.

Delwiche, Aaron Alan, and Jennifer Jacobs Henderson. "Introduction: What Is Participatory Culture?" In *The Participatory Cultures Handbook*, edited by Jennifer Jacobs Henderson and Aaron Alan Delwiche, 3–11. Routledge, 2013.

Doerschuk, Bob. "Orchestral Manoeuvres in the Dark." *Keyboard*, April 1982, 30.
Dolan, Emily I. *The Orchestral Revolution: Haydn and the Technologies of Timbre*. Cambridge University Press, 2013.
Dolan, Emily I., and John Tresch. "A Sublime Invasion: Meyerbeer, Balzac, and the Opera Machine." *Opera Quarterly* 27, no. 1 (2011): 4–31.
Doran, John. "Karl Bartos Interviewed: Kraftwerk and the Birth of the Modern." *The Quietus*, March 11, 2009. https://thequietus.com/articles/01282-karl-bartos-interviewed-kraftwerk-and-the-birth-of-the-modern.
Duffett, Mark. "Average White Band: Kraftwerk and the Politics of Race." In *Kraftwerk: Music Non-Stop*, edited by Sean Albiez and David Pattie, 194–213. Continuum, 2011.
Duguid, Paul. "Material Matters: The Past and Futurology of the Book." In *The Future of the Book*, edited by Geoffrey Nunberg, 63–101. University of California Press, 1996.
Dwango Co. "NicoNico Chokaigi." 2019. https://chokaigi.jp/2019/en/about.
Dyson, T. G. "Mechanical Pianoforte Players." *Musical Opinion & Music Trade Review* 25, no. 291 (December 1901): 231.
"Editor's Overview." *Modern Drummer*, February/March 1982, 2.
Ehrlich, Cyril. *The Piano: A History*. Oxford University Press, 1990.
Eidsheim, Nina Sun. "Race as Zeroes and Ones: Vocaloid Refused, Reimagined, and Repurposed." In *The Race of Sound: Listening, Timbre, and Vocality in African-American Music*, 115–50. Duke University Press, 2019.
Eyerman, Charlotte N. "Piano Playing in Nineteenth-Century French Visual Culture." In *Piano Roles: A New History of the Piano*, edited by James Parakilas, 180–202. Yale University Press, 2001.
"First International Sewing Machine and Domestic Appliances Exhibition." *Journal of Domestic Appliances and Sewing Machine Gazette*, December 1, 1887, 24–25.
Flans, Robyn. "Classic Tracks: Phil Collins' 'In the Air Tonight.'" *Mix*, May 1, 2005. https://www.mixonline.com/mag/audio_phil_collins_air/index.html.
Flans, Robyn. "Harvey Mason." *Modern Drummer*, July 1981, 64.
Flans, Robyn. "Jeff Porcaro." *Modern Drummer*, February/March 1982, 19.
Flans, Robyn. "Jeff Porcaro." *Modern Drummer*, February 1983, 46.
Flans, Robyn. "Jim Keltner." *Modern Drummer*, February/March 1982, 19–20.
Flans, Robyn. "Narada Michael Walden: Inspired." *Modern Drummer*, April 1982, 29, 86.
Flür, Wolfgang, Janet Porteous, and Barbara Uhling. *Kraftwerk: I Was a Robot*. Omnibus Press, 2017.
Gann, Kyle. *The Music of Conlon Nancarrow*. Cambridge University Press, 1995.
Garcia, Cory. "Hatsune Miku's Hologram Concert Is a Sure Sign of Endtimes, and I Can't Wait." *Dallas Observer*, November 17, 2015. https://www.dallasobserver.com/music/hatsune-mikus-hologram-concert-is-a-sure-sign-of-endtimes-and-i-cant-wait-7780679.

Gay, Leslie C., Jr. "Before the Deluge: The Technoculture of Song-Sheet Publishing Viewed from Late-Nineteenth-Century Galveston." In *Music and Technoculture*, edited by Rene T. A. Lysloff and Leslie C. Gay Jr., 208–32. Wesleyan University Press, 2003.

Geertz, Clifford. "Religion as a Cultural System." In *The Interpretation of Cultures*, 87–125. Basic Books, 1973.

Geo. J. Birkel Co. Advertisement. *Los Angeles Times*, November 21, 1909, II.

Gitelman, Lisa. "Media, Materiality, and the Measure of the Digital; or, The Case of Sheet Music and the Problem of Piano Rolls." In *Memory Bytes: History, Technology, and Digital Culture*, edited by Lauren Rabinovitz and Abraham Geil, 199–217. Duke University Press, 2004.

Gitelman, Lisa, and Geoffrey Pingree. "What's New About New Media?" In *New Media, 1740–1915*, xi–xxi. MIT Press, 2003.

Givens, Larry. *Re-Enacting the Artist: The Story of the Ampico Reproducing Piano*. The Vestal Press, 1970.

Goto, Masataka. "Frontiers of Music Technologies: Singing Synthesis and Active Music Listening." Presentation at the CIRMMT Distinguished Lecture, Montreal, QC, September 24, 2015.

Goto, Masataka. "OngaCREST Project: Building a Similarity-Aware Information Environment for a Content-Symbiotic Society." In *Human-Harmonized Information Technology, Volume 2, Horizontal Expansion*, edited by Toyoaki Nishida, 1–3. Springer Japan, 2017.

Grant, Steven. "Is Gary Numan Eclectic?" *Trouser Press*, May 1980, 6.

Greenall, John. Advertisement. *Journal of Domestic Appliances and Sewing Machine Gazette*, January 1, 1886, 33.

Greene, Andy. "Phil Collins: My Life in 15 Songs." *Rolling Stone*, February 29, 2016. https://www.rollingstone.com/music/music-lists/phil-collins-my-life-in-15-songs-82641.

Grew, Sydney. *The Art of the Player-Piano: A Textbook for Student and Teacher*. Kegan Paul, Trench, Trubner & Co., 1922.

Gronholm, Johannes. "Kraftwerk." In *Kraftwerk: Music Non-Stop*, edited by Sean Albiez and David Pattie, 77–94. Continuum, 2011.

Hadfield, James. "YMO's Yukihiro Takahashi Celebrates a Pair of 40th Anniversaries." *Japan Times*, November 23, 2018. https://www.japantimes.co.jp/culture/2018/11/23/music/ymos-yukihiro-takahashi-celebrates-pair-40th-anniversaries/#.XQhczBZKhaQ.

Hall, Stanley. "Simon Phillips." *Modern Drummer*, June 1981, 11.

Hankins, Thomas L., and Robert J. Silverman. *Instruments and Imagination*. Princeton University Press, 2014.

Haraway, Donna J. *Simians, Cyborgs, and Women: The Reinvention of Nature*. Routledge, 1991.

Hardman, Peck & Co. Advertisement. *New York Times*, October 11, 1925.

Harris, Neil. *Cultural Excursions: Marketing Appetites and Cultural Tastes in Modern America*. University of Chicago Press, 1990.

Hasse Jørgensen, Stina, Sabrina Vitting-Seerup, and Katrine Wallevik. "Hatsune Miku: An Uncertain Image." *Digital Creativity* 28, no. 4 (2017): 318–31.

Hatsune miku ga orijinaru kyoku o utatte kureta yo (初音ミク が オリジナル曲を歌ってくれたよ「メルト」) [Hatsune Miku sang original song "Melt"]. *NicoNico*. Posted December 7, 2007, by Ryo. https://www.nicovideo.jp/watch/sm1715919.

Hatsune miku odds & ends PV full version (【初音ミク】 ODDS & ENDS—PV Full Ver.) [Hatsune Miku ODDS & ENDS—PV Full Ver.]. *NicoNico*. Posted August 12, 2012, by まさゆき. https://www.nicovideo.jp/watch/sm18592204.

Hatsune miku orijinaru kyoku "hatsune miku no shoushitsu (long version)" (初音ミクオリジナル曲「初音ミクの消失 [LONG VERSION]) [An original Hatsune Miku song 'The Disappearance of Hatsune Miku (long version)]. *NicoNico*. Posted April 8, 2008, by cosMo@暴走P. https://www.nicovideo.jp/watch/sm2937784.

Hawley, Michelle. "Apple's 'Crush' Ad Controversy: Insights into the Power of Public Opinion." *CMSWire*, May 13, 2024. https://www.cmswire.com/digital-marketing/apples-crush-ad-controversy-insights-into-the-power-of-public-opinion/.

Henke, James. "Yellow Magic Orchestra: Japanese Technopop Is Poised to Invade America." *Rolling Stone*, no. 319, June 12, 1980.

Hickman, C. N. "A Spark Chronograph Developed for Measuring Loudness of Piano Tones." *Journal of the Acoustical Society of America* 1, no. 35 (1929): 138–46.

"Hire Agreements—Important." *Journal of Domestic Appliances and Sewing Machine Gazette*, July 1, 1886, 13.

Holliday, Kent A. *Reproducing Pianos Past and Present*. Edwin Mellen Press, 1989.

Holliday, Kent. "Some American Firms and Their Contributions to the Development of the Reproducing Piano." In *Perspectives on American Music, 1900–1950*, edited by Michael Saffle, 97–122. Routledge, 2012.

Howe, Sondra Wieland. "Music in the Private Sphere, Churches, and Community." In *Women Music Educators in the United States: A History*, 43–73. Scarecrow Press, 2014.

Hütter, Ralf. "Interview by Beacon Radio." Birmingham, UK, June 14, 1981. https://archive.org/web/20081202043525/http://kraftwerk.technopop.com.br/interview_122.php.

"Hydraulic Press Channel." *SocialBlade*. Accessed April 14, 2025. https://socialblade.com/youtube/channel/UCcMDMoNu66_1Hwi5-MeiQgw/monthly.

Inaba, Minoru. "Computer Rock Music Gaining Fans." *Sarasota Journal*, August 18, 1980, 11A.

Jenkins, Henry. *Confronting the Challenges of Participatory Culture: Media Education for the 21st Century*. MIT Press, 2009.

Jenkins, Henry. *Fans, Bloggers, and Gamers: Exploring Participatory Culture*. New York University Press, 2006.

Johnstone, Rob. *Kraftwerk and the Electronic Revolution*. Films Media Group, 2013. eVideo.

Joseph, Charles M. *Stravinsky's Ballets*. Yale University Press, 2011.

Kahney, Leander. "Apple's 'Crush' iPad Ad Draws Fire." *Cult of Mac Newsletter*, May 8, 2024. https://newsletters.cultofmac.com/p/apples-crush-ipad-ad-draws-fire.

Kane, Carolyn L. *High-Tech Trash: Glitch, Noise, and Aesthetic Failure*. University of California Press, 2019.

Katz, Mark. "The Amateur in the Age of Mechanical Music." In *The Oxford Handbook of Sound Studies*, edited by Trevor Pinch and Karin Bijsterveld, 105–25. Oxford University Press, 2012.

Keightley, Keir. "Reconsidering Rock." In *The Cambridge Companion to Pop and Rock*, edited by Simon Frith, Will Straw, and John Street, 109–42. Cambridge University Press, 2001.

Keightley, Keir. "'Turn It Down!' She Shrieked: Gender, Domestic Space, and High Fidelity, 1948–59." *Popular Music* 15 (1996): 149–77.

Kempelen, Wolfgang von. *Mechanismus der menschlichen Sprache nebst der Beschreibung einer sprechenden Maschine*. J. B. Degen, 1791. Quoted in Willem Levelt, *A History of Psycholinguistics: The Pre-Chomskyan Era*, 129–31. Oxford University Press, 2013.

"Kempelen's Speaking Machine." Posted June 6, 2017, by Fabian Brackhane. YouTube. https://www.youtube.com/watch?v=k_YUB_S6Gpo.

Kenner, Hugh. *The Counterfeiters: An Historical Comedy*. Dalkey Archive Press, 2005.

Klatt, Dennis H. "Review of Text-to-Speech Conversion for English." *Journal of the Acoustical Society of America* 82 (1987): 737–93.

Knabe. Advertisement. *New York Times*, April 14, 1920, 7.

Knabe. Advertisement. *New York Times*, January 15, 1920, 16.

Knabe. Advertisement. *New York Tribune*, March 24, 1918, F10.

Kranich & Bach. Advertisement. *American Homes and Gardens*, September 1912, back cover.

Landay Bros. Advertisement. *New York Times*, July 27, 1908, 2.

Landay Bros. Advertisement. *New York Times*, December 24, 1908, 2.

Lawson, Rex. "On the Right Track—Dynamic Recording for the Reproducing Piano (Part Four)." *Pianola Journal* 23 (2013): 11–21.

Lawson, Rex. "On the Right Track: The Recording of Dynamics for the Reproducing Piano (Part One)." *Pianola Journal* 20 (2009): 6–11.

Lerner, Neil. "Mario's Dynamic Leaps: Musical Innovations (and the Specter of Early Cinema) in Donkey Kong and Super Mario Bros." In *Music in Video Games: Studying Play*, edited by K. J. Donnelly, William Gibbons, and Neil Lerner, 1–16. Routledge, 2014.

LeRoy, Dan. *Dancing to the Drum Machine*. Bloomsbury Academic, 2023.

Levelt, Willem. *A History of Psycholinguistics: The Pre-Chomskyan Era.* Oxford University Press, 2013.

Littlejohn, John T. "Kraftwerk: Language, Lucre, and Loss of Identity." *Popular Music and Society* 32 (2009): 635–53.

Loesser, Arthur. *Men, Women, and Pianos.* Simon and Schuster, 1954.

Lucas-Flannery, Mary. "What I Learned from a Vocaloid and 1500 Screaming Millennials." *Medium,* May 5, 2016. https://medium.com/@marylucusflannery/what-i-learned-from-a-vocaloid-and-1500-screaming-millennials-d181a26d5ce6.

Macias, Patrick. "CRN Interview: The Creators of Hatsune Miku." *Crunchyroll,* July 21, 2011. https://www.crunchyroll.com/anime-feature/2011/07/21/crn-interview-the-creators-of-hatsune-miku.

Manning, Peter. *Electronic and Computer Music.* Oxford University Press, 2013.

Mansfield, Joe. *Beat Box: A Drum Machine Obsession.* Get On Down, 2013.

Marchand, Roland. *Advertising the American Dream: Making Way for Modernity, 1920–1940.* University of California Press, 1986.

Marsh, Calum. "We Attended the Hatsune Miku Expo to Find Out If a Hologram Pop Star Could Be Human." *Vice,* May 27, 2016. https://www.vice.com/en_us/article/ae87yb/hatsune-miku-expo-feature.

Mattingly, Rick. "Mel Lewis." *Modern Drummer,* February 1985, 8, 50.

Mattingly, Rick. "Roger Linn." *Modern Drummer,* February/March 1982, 20, 100.

Mattingly, Rick. "Terry Bozzio." *Modern Drummer,* December 1984, 60.

McEvoy, J. P. "The Player-Piano Upstairs." In *Slams of Life, with Malice for All and Charity Toward None,* 22–23. P. F. Volland, 1919.

McGovern, Charles. *Sold American: Consumption and Citizenship, 1890–1945.* University of North Carolina Press, 2006.

Meyer-Frazier, Petra. "Music, Novels, and Women: Nineteenth-Century Prescriptions for an Ideal Life." *Women and Music: A Journal of Gender and Culture* 10 (2006): 45–46.

Michler, Robert. "A Romanticized Narrative and the Overlooked Birth of Electronic Beats." In *Studies in the Arts II—Künste, Design und Wissenschaft im Austausch,* edited by Thomas Gartmann, Cristina Urchueguía, and Hannah Ambühl-Baur. Transcript Verlag, 2023.

Mickle, Tripp. "Apple Says Destructive iPad Ad 'Missed the Mark.'" *New York Times,* May 9, 2024, https://www.nytimes.com/2024/05/09/technology/apple-ipad-ad-crush-apology.html.

"A Momentous Musical Meeting." *Etude,* October 1923, 663.

Morris, Chris. "Wendy Carlos Takes Her Moog Music to East Side Digital." *Billboard,* October 3, 1998, 69.

National Commission on Technology, Automation, and Economic Progress. *Hearings Before the Select Subcommittee on Labor, of the Committee on Education and Labor, House of Representatives, Eighty-Eighth Congress, Second Session, on H.R. 10310, and Related Bills to Establish a National*

Commission on Automation and Technological Progress. US Government Printing Office, 1964.

NBT and Sizergyia. "Interview with the Creator of Hatsune Miku, Hiroyuki Itoh." *JRock News*, March 11, 2019. https://jrocknews.com/2019/03/interview-hiroyuki-itoh-creator-of-hatsune-miku.html.

Neves, M. J. "The Dehumanization of Art: Ortega y Gasset's Vision of New Music." *International Review of the Aesthetics and Sociology of Music* 43 (December 2012): 363–78.

Newman, Ernest. *The Piano-Player and Its Music*. Riverside Press, 1920.

"A Notable Presentation of a Notable Instrument." *New York Tribune*, November 25, 1917, 6.

Nye, Sean. "Minimal Understandings: The Berlin Decade, the Minimal Continuum, and Debates on the Legacy of German Techno." *Journal of Popular Music Studies* 25 (June 2013): 154–84.

Ord-Hume, Arthur W. J. G. *Pianola: The History of the Self-Playing Piano*. Allen & Unwin, 1984.

Oricon News. "Supercell のメンバーにインタビュー" [Interview with Supercell Members]. Accessed December 11, 2018. https://www.oricon.co.jp/trend/hayari/20080805_02.html.

Patteson, Thomas. *Instruments for New Music: Sound, Technology, and Modernism*. University of California Press, 2016.

Pattie, David. "Introduction: The (Ger)Man Machines." In *Kraftwerk: Music Non-stop*, edited by Sean Albiez and David Pattie, 9. Continuum, 2011.

Pattie, David. "Kraftwerk: Playing the Machines." In *Kraftwerk: Music Non-Stop*, edited by Sean Albiez and David Pattie, 115–27. Continuum, 2011.

Pesic, Peter. "Music of the Clocks and Spheres: Mozart and Haydn's Experiments with Time." *Eighteenth-Century Music* 21, no. 2 (September 2024): 157–85.

Peters, John Durham. *Speaking into the Air: A History of the Idea of Communication*. University of Chicago Press, 1999.

"The Piano Player Vogue." *Musical Opinion & Music Trade Review* 26, no. 304 (January 1903): 309.

Pinch, Trevor J., and Wiebe E. Bijker. "The Social Construction of Facts and Artefacts: Or How the Sociology of Science and the Sociology of Technology Might Benefit Each Other." *Social Studies of Science* 14 (August 1984): 399–441.

Pinch, Trevor, and Frank Trocco. *Analog Days: The Invention and Impact of the Moog Synthesizer*. Harvard University Press, 2002.

"Player-Piano Nomenclature: Words to Use and Avoid." *Player Piano* 1, no. 5 (September 1911): 5–6.

Q. "Q Review: Enter Hatsune Miku's Hologram Concert." *CBC Radio*, May 24, 2016. https://www.cbc.ca/radio/q/schedule-for-tuesday-may-24-2016-1.3597178/q-review-enter-hatsune-miku-s-hologram-concert-1.3597187.

Reason, Matthew, and Anja Mølle Lindelof. *Experiencing Liveness in Contemporary Performance: Interdisciplinary Perspectives*. Routledge, 2014.

Renfro, Kim. "Here's Why Modern Songs Play on the Saloon's Piano in *Westworld*." *Insider*, October 11, 2016. https://thisisinsider.com/westworld-piano-songs-2016-10.

Richards, Annette. "Automatic Genius: Mozart and the Mechanical Sublime." *Music & Letters* 80, no. 3 (August 1999): 366–89.

Riskin, Jessica. "The Defecating Duck, or, the Ambiguous Origins of Artificial Life." *Critical Inquiry* 29 (Summer 2003): 599–609.

Roehl, Harvey. *Player Piano Treasury*. 2nd ed. Vestal Press, 1973.

Roehl, Harvey. *Player Piano Treasury: The Scrapbook History of the Mechanical Piano in America, As Told in Story, Pictures, Trade Journal Articles and Advertising*. Vestal Press, 1961.

"Roger Linn: Technical GRAMMY Award Acceptance." *Grammy Awards*, February 13, 2011. https://www.grammy.com/videos/roger-linn-at-special-merit-awards-ceremony-nominees-reception.

Rousseau, Jean-Jacques. "Essay on the Origin of Language." In *On the Origin of Language*, translated by John H. Moran and Alexander Gode, 1–74. University of Chicago Press, 1966.

Rushkoff, Douglas. *Present Shock: When Everything Happens Now*. Penguin Group, 2013.

Sanden, Paul. "Virtual Liveness and Sounding Cyborgs: John Oswald's 'Vane.'" *Popular Music* 31, no. 1 (2012): 45–68.

Savage, Jon. "Interview: Kraftwerk." Red Bull Music Academy, August 30, 2012. https://daily.redbullmusicacademy.com/2012/08/kraftwerk-interview.

Schaefer, John. "Shuggie Otis Spreads His 'Wings,' 40 Years Later." New Sounds. April 16, 2013. https://newsounds.org/story/287163-shuggie-otis-wings.

Schedel, Margaret. "Anticipating Interactivity: Henry Cowell and the Rhythmicon." *Organised Sound* 7, no. 3 (December 2002): 247–54.

Schneider, Mitchell. "The Man-Machine." *Rolling Stone*, May 18, 1978. https://www.rollingstone.com/music/music-album-reviews/the-man-machine-96960.

Seaver, Nick. "'This Is Not a Copy': Mechanical Fidelity and the Re-enacting Piano." *Differences* 22, nos. 2–3 (2011): 56–84.

Sigman, Alexander. "Robot Opera: Bridging the Anthropocentric and the Mechanized Eccentric." *Computer Music Journal* 43, no. 1 (2019): 21–37.

Simonton, Richard C., and Ben Hall. Liner Notes to *The Welte Legacy of Piano Treasures*. Hathaway and Bowers, HSTR-5370, 1963. LP.

Slade, Alison, Amber J. Narro, and Dedria Givens-Carroll, eds. *Television, Social Media, and Fan Culture*. Lexington Books, 2015.

Slonimsky, Nicolas. "Henry Cowell." In *American Composers on American Music: A Symposium*, edited by Henry Cowell, 59–60. Frederick Ungar, 1962.

Smith, Bertram. "The Piano-Player." *Musical Times* 52 (May 1, 1911): 308–9. Accessed August 4, 2024. https://songrium.jp.

Sousa, John Philip. "The Menace of Mechanical Music." *Appleton's Magazine* 8 (August 1906): 278–84.

Sousa, John Philip. *New York Morning Telegraph*, June 12, 1906. Quoted in Neil Harris, *Cultural Excursions: Marketing Appetites and Cultural Tastes in Modern America*, 224. University of Chicago Press, 1990.
Steinway & Sons. Advertisement. *Country Life* 31, no. 803 (May 25, 1912): lxxxix.
Stilwell, Robynn J. "In the Air Tonight: Text, Intertextuality, and the Construction of Meaning." *Popular Music and Society* 19, no. 4 (1995): 93–97.
Suisman, David. "Sound, Knowledge, and the 'Immanence of Human Failure': Rethinking Musical Mechanization Through the Phonograph, the Player-Piano, and the Piano." *Social Text* 28 (Spring 2010): 13–14.
Sumrall, Harry. "Yellow Magic." *Washington Post*, November 6, 1979, B7.
Supercell. "Biography." Accessed July 8, 2019. https://supercell.jp/biography.html.
Supercell. "Discography." Accessed July 8, 2019. https://supercell.jp/discography.html.
"Symphony Players in Novel Concert." *New York Sun*, November 18, 1917, 9.
Taruskin, Richard. *Stravinsky and the Russian Traditions: A Biography of the Works Through Mavra*. University of California Press, 1996.
Taylor, Timothy. "The Commodification of Music at the Dawn of the Era of 'Mechanical Music.'" *Ethnomusicology* 51 (Spring/Summer 2007): 281–305.
Thompson, Emily. "Machines, Music, and the Quest for Fidelity: Marketing the Edison Phonograph in America, 1877–1925." *Musical Quarterly* 79 (1995): 159–60.
Thomson, Ross. *Mechanized Shoe Production in the United States: The Rise of the Modern Manufacturing System*. Cambridge University Press, 1989.
Toffler, Alvin. *Future Shock*. Random House, 1970.
Toffler, Alvin. *The Third Wave*. Bantam Books, 1980.
"The Triumph Piano Player." *Musical Opinion & Music Trade Review* 27, no. 318 (March 1904): 483.
Turkle, Sherry. *Alone Together: Why We Expect More from Technology and Less from Each Other*. Basic Books, 2011.
United States Department of Labor. *War and Postwar Wages, Prices, and Hours*. US Government Printing Office, 1945.
Velasquez, Tito. Interview by John Glanville, Research Assistant. Copenhagen Virtual Concert Party Organizer, May 21, 2023.
Warfield, Patrick. "John Philip Sousa and 'The Menace of Mechanical Music.'" *Journal of the Society for American Music* 3, no. 4 (2009): 431–63.
Weinstein, Leon. "Hatsune Miku at the Sony Centre for the Performing Arts." *Live in Limbo*, May 23, 2016. https://liveinlimbo.com/2016/05/23/concert-reviews/hatsune-at-the-sony-centre-for-the-performing-arts.html.
The Welte-Mignon Autograph Piano Co. Advertisement. *New York Times*, December 7, 1911, 13.
Wente, Allison Rebecca. *The Player Piano and Musical Labor*. Routledge, 2022.
Wetherbee, Brandon. "Live DC: Hatsune Miku Expo @ The Anthem." *Brightest Young Things*, July 13, 2018. https://brightestyoungthings.com/articles/live-dc-hatsune-miku-expo-the-anthem.

Wilcox & White Co. Advertisement. *New York Times*, May 1, 1898, 20.
Wilkinson, Scott. "Humanoid or Vocaloid?" *Electronic Musician*, August 1, 2003. https://www.emusician.com/mag/emusic_humanoid_vocaloid/index.html.
Wurlitzer. Advertisement. *International Musician* 59, no. 8 (February 1961): 6.
Yamada, Keisuke. *Supercell Featuring Hatsune Miku*. Bloomsbury Academic, 2017.
Yano, Christine R. "Letters from the Heart: Negotiating Fan-Star Relationships in Japanese Popular Music." In *Fanning the Flames: Fans and Consumer Culture in Contemporary Japan*, edited by William W. Kelly, 41–58. SUNY Press, 2004.
"Yellow Magic Orch." *Variety*, December 17, 1980, 72.
Yin, Yiyi. "Vocaloid in China: Cosmopolitan Music, Cultural Expression, and Multilayer Identity." *Global Media and China* 3, no. 1 (2018): 51–66.
Zaborowski, Rafal. "Hatsune Miku and Japanese Virtual Idols." In *The Oxford Handbook of Music and Virtuality*, edited by Sheila Whiteley and Shara Rambarran, 121–29. Oxford University Press, 2016.

Index

Abbate, Carolyn, 35
ABBA Voyage, 212n17
Adelt, Ulrich, 84–85
Advanced Industrial Science and Technology (AIST), 145
advertisements, 14–15, 16, 20, 26, 61–62, 63, 76, 189–90; for the "automatic piano," 29–30; for hi-fi systems, 202–3n47; for the phonograph, 55. *See also* Pianola, first advertisements for; player piano, the, in early advertisements and articles; reproducing pianos, advertisements for; Welte-Mignon reproducing piano, advertisements for
Aeolian, 30, 52, 57, 76–77, 201n23. *See also* Duo-Art
AI (artificial intelligence), 6, 188
Akai DPS12, 137
"All I Need Is You" (Sonny and Cher), 187
American Federation of Musicians (AFM), 12, 113
American Homes and Gardens, 42
American Piano Company, 56–57, 65, 70
American Pneumatic Service Company, 65
American Society of Composers, Authors and Publishers (ASCAP), 195

Ampico, 56, 57–59, 62; early years of, 62–67; prices of the grand pianos produced by, 58; spark chronograph of, 68–74, 70*fig.*; woodworking department of, 69*fig*. *See also* "Cookie Chronograph"
Ampico Bulletin, 56
Ampico Re-Enacting Piano, 65
Ampico *Service Manual*, 68
Angelus Orchestral Piano Player, 22–23, 23*fig.*, 30, 31, 31*fig.*
Anti-Pianola, 38–39
anxiety. *See* technological anxiety
Apple iPad Pro, promotional video for ("Crush"), 187–90; backlash against, 196
Archies, The, 88
Art of the Player-Piano, The: A Text-Book for Student and Teacher (Grew), 43–44
artistic/mechanical, artificial distinction between, 8
audience/performance relationship, 2
authenticity, 190; and German automata, 84–92; in rock music, 87
automata, 34–35, 42, 88, 93–94, 111. *See also* authenticity, and German automata; robots
Automata (2014), 13
automation/automation technology, 3, 6, 12, 16, 18, 20–21, 31, 62, 90, 127,

227

228 / Index

automation/automation technology (*continued*) 191; conceptual instability of, 159; cultural history of, 3; debates concerning automation technology in music, 7; difference of from mechanization, 9–10; fear and resentment of concerning the development of, 190–93; and musical performance, 9; reception and use of, 14–16; and repeatability, 185; three categories of, 3–4. *See also* innovation, human and technological; new music/new music technologies; technological anxiety; technological optimism

avatar performances, 4, 5. *See also* Hatsune Miku; NicoNico Cho Party

Babbitt, Milton, 83
Bangs, Lester, 84–92
Barker, Martin, 170–71; "experiential excess" concept of, 179–80
Bartos, Karl, 91, 207n33
Bauer, Harold, 51–52
Beatles, The, 103
Berry, Chuck, 87
Black Mirror (TV series [2011–]), 13
Blade Runner (1982), 12; Vangelis's music in, 12
Blade Runner 2049 (2017), 13
Bleach anime, 210n15
Bohemian Dance (Smetana), 46
Bon Iver, 13
Bourdaghs, Michael K., 104
Bozzio, Terry, 125–26
Brady, Cyrus Townsend, 59–60
Braine, Robert, 32–34, 38, 40
Braverman, Harry, 28
British Musicians' Union, 12
Butler, Mark, 10

Callas, Maria, 159, 163; hologram concert/performance of, 2
Carlos, Wendy, 12, 80
Carr, Robert, 118
Carreno, Teresa, 61

Caswell, Ken, 64
Cateforis, Theo, 91
Circus, 103–4
Clapton, Eric, 106
Coachella, 163
Collins, Phil, 129, 130–131
Complete Organist, The, 41
CompuRhythm CR-78, 114, 130
"Computer Game: 'Theme from The Circus'" (YMO), 103–4
concert formats, contrasting, 184–86
"Cookie Chronograph," 67
Cornelius, Don, 99–101
Cowell, Henry, 111–12
creativity, human, 5, 7; and agency, 40–43. *See also* innovation, human and technological
crisis historiography, 14
Crypton Future Media, 158–59
Cummings, William H., 39–40
Currid, Brian, 104
cymbal sounds, decay of, 114

"Daisy Bell" (IBM 7094 program of tune), 133–34, 210n5
Damrosch, Walter, 51
Davis, Ben, 37
"Day Tripper" (The Beatles), 105; YMO's version of, 105–6
Death Cab for Cutie, 132
dehumanization, 6, 10, 40, 86–87, 91, 107, 108, 127
Delasaire, William, 46
Delwiche, Aaron, 147–48
democratization, 192, 196
Denny, Martin, 103
derivative works, in a content-symbiotic society, 145–49
desktop music (DTM), 107; culture of, 136–37
"Die Mensch-Maschine: Kraftwerk," 197
digital performances, 1–2. *See also* avatar performances
"Disappearance of Hatsune Miku -DEAD END-, The" (cosMo), 182, 183*fig.*

disco, 11, 97, 127–128
DJs, 106
Dolan, Emily, 8, 35
"Don't You Want Me" (Human League), 129
drummers, 109–11, 113, 115, 124–29, 131, 132, 190; jazz drummers, 123. *See also* drum machines; drums/drum kits; Lewis, Mel, on drum machines; *Modern Drummer*, discussion in by professional drummers concerning drum machines
drum machines, 4, 5, 8–9, 11, 17, 108–11, 116–17, 208n7; arrival of in popular music, 111–15; creativity and the drum machine (1981–1985), 128–31. *See also* Lewis, Mel, on drum machines; Linn Drum; LM-1 drum machine; *Modern Drummer*, discussion in by professional drummers concerning drum machines
drums/drum kits, 8, 111; and double-kick pedals, 8; positioning of, 111; and synthetic drum heads, 8
Dudley, Homer, 133
Duo-Art, 51–52, 57, 74–77
DX synthesizer series, 137–38; and the DX7, 137; and the DX100, 138
Dyson, T. G., 38, 40

East Asia, 102, 164
"edge cases," 194
Edison, Thomas, 195
Eiffel Tower, 10
Electronics & Music Maker, 90
electronic dance music (EDM), 106, 193
END, THE (opera), 193
Etude, The, 27, 32–33
Ex Machina (2014), 13

Fairchild, Edgar, 66
"fidelity," socially constructed concept of, 79
"Firecracker" (Denny), 102

Flans, Robyn, 118–21, 129, 130
Flür, Wolfgang, 83, 93, 207n33
Forster, E. M., 10
Future Eve, The (l'Isle-Adam), 10

Geertz, Clifford, 165
Givens, Larry, 66
Godowsky, Mr., 59, 61
Gorillaz, 159
Goto, Masataka, 145, 146, 158
Grainger, Percy, 74–77, 75*fig.*, 79
Greenall Steam Washer, advertisement for, 27
Grew, Sydney, 43–46, 48; and good pedaling, 45; systematic approach to musicianship and technical competence, 44
Gronholm, Johannes, 207n33

hammered dulcimer, 9
hammer velocity/human performance, subjective relationship between, 72–73
Handel, George Frideric, 40
Haraway, Donna, 196
hard drives, affordability of, 137
Hatsune Miku, 2, 6, 134–35, 135*fig.*, 137–38, 139, 140, 141–44, 150–52, 152*fig.*, 156–61, 166, 169, 193; concerts of, 6, 160, 167; criticism of, 158; fans of, 4; first appearance of in the West, 212n10; function and identity of in Supercell, 153; as a hologram, 4; popularity of, 159; reactions to performances of, 1. *See also* Miku Expo tours
"Hatsune Miku no Shoushitsu -DEAD END-" (cosMo), 182, 183*fig.*
Haydn, Joseph, 34
Henderson, Jennifer Jacobs, 147–48
Her (2013), 13
Hickman, Clarence, 68–69, 71
hi-fi systems, advertising for, 202–3n47
Holliday, Kent A., 64
holographic bodies/human bodies, 163–65

holographic performances, 2, 16, 91; ethics of, 13; framing of a holographic concert, 165–69
home recording, 13
Hosono, Haruomi, 96, 97
Houston, Whitney, 157
Hütter, Ralf, 80, 83, 84–85, 89, 90–91, 197–98
Human League, 129, 130, 131

IBM 7094, 133–34, 210n5
ingenuity, human. *See* innovation, human and technological
"I'll Miku Miku You" (ika), 143–44, 146–47, 147*fig.*
"In the Air Tonight" (Collins), 129, 130
innovation, human and technological, 3, 5, 7, 10, 34, 39, 66, 111, 116, 165, 192, 193, 196
Intel, 11
International Musician, 113
International Sewing Machine and Domestic Appliances Exhibition, 29
Ito, Youichi, 101

Jackson, Michael, 106, 126
Jenkins, Henry, 146
John, Elton, 108
Journal of the Acoustical Society of America, 71
Journal of Domestic Appliances and Sewing Machine Gazette, 27–28, 29
journalists, 15

Karlin, Fred, 12
Keightley, Keir, 87, 104
Kelly, John, 133–34
Keltner, Jim, 118–19, 121–23, 128
kick-pedals, 8; double-kick pedals, 8
Kiss, 88
"Koisuru VOC@LOID" (OSTER project), 143
Kool & the Gang, 101
Kraftwerk, 11, 17, 80–81, 98, 100, 103, 105, 131, 136, 197, 198; *Autobahn* album of, 85–86, 104; *Autobahn* tour of, 84, 96; international success of, 82–84; "man-machine" concept of, 86, 90, 94; popularity of, 107; rejection of 1970s rock ideals, 91; robotic performance aesthetics of, 92–96, 95*fig.*; synthesizers used by, 206n11
"Kraftwerk—Die Mensch Maschine," 86
Kranich & Bach, 42

Lady Gaga, 158
laptop performers, 106
Lawson, Rex, 48–49, 67, 192, 203n60
Lewis, Mel, on drum machines, 123–24, 128, 132, 190
Life of Richard Wagner, The (Newman), 41
light/glow sticks, 164–65, 172*fig.*, 179*fig.*, 211n1
Linn, Roger, 108, 109, 114, 115, 130, 131–32, 192–93
LinnDrum, 124, 126
l'Isle-Adam, Villiers de, 10
LM-1 drum machine, 17, 108, 109, 110, 114–15, 117, 118, 129–30, 132; advertisement for, 109*fig.*
Lockbaum, Carol, 133–34
Loughridge, Deirdre, 199n8
Luka, Megurine, 173
"Luka Luka Night Fever," 173

M. Welte und Söhne, 25–26
"Machine Stops, The" (Forster), 10
Maestro Rhythm King, 114
Man-Machine, The (Kraftwerk), 83, 92
Mason, Harvey, 117
Matsutake, Hideki, 99
Mattingly, Rick, 108, 125–26
McGerr, Jason, 132
"mechanics," 55, 111
mechanization, 9, 28; difference of from automation, 9–10
"Melt" (Ryo), 150
Memorising Music (Newman), 41
"Menace of Mechanical Music" (Sousa), 36, 37*fig.*, 194
Mendelssohn, Felix, 37–38

Merlin, Joseph, 56
Michler, Robert, 208n7
MIDI (Musical Instrument Digital Interface), 136
Miku. *See* Hatsune Miku
Miku Expo tours (2016 and 2018), 160, 161–62, 168; fan participation at (2016), 163–65; Miku Expo2018, 169–77, 172*fig*., 174*fig*., 175*fig*.
MikuMikuDance (MMD), 148–49
"Miku Miku ni Shite Ageru" (ika), 143–44, 146–47, 147*fig*.
Missing Persons, 125
Mix, 130
Modern Drummer, 108, 109; discussion in by professional drummers concerning drum machines, 116–23; beyond the 1982 forum concerning (the drum machine as a creative tool), 123–29
Moffett, Alex, 114
Monkees, The, 88
Moog, Robert, 83, 142
Mozart, Wolfgang, 34–35
Musical Courier, 27
musical data, 137
musical identity, 83–84
Musical Opinion & Music Trade Review, 35–36, 38
musical performance, definitions of, 11
Musical and Sewing Machine Courier, 27
Musical and Sewing Machine Gazette, 27
Musical Times, 26, 39, 45, 46
music producers, 127–28
Music Star Map, 147*fig*.
music technologies, 7, 8. *See also* new music/new music technologies
music videos, animated, 148
"Musique Non Stop" (Kraftwerk), 198
MySpace, 13

Nancarrow, Conlon, 21, 49, 112
Newman, Ernest, 28–29, 41–42, 43
new music/new music technologies, 4, 6, 20–21, 87, 110, 117, 118, 124, 125, 129, 160; in Japan, 205–6n3

New Musical Express, 85
New Musical Resources, 112
New York Symphony Orchestra, 51, 52*fig*.
NicoNico Chokaigi, 160; Vocaloid music and audience participation at, 177–80, 179*fig*.
NicoNico Cho Party, 160, 178, 180–84, 183*fig*.
NicoNico Douga, 138–45, 141*fig*., 144–45, 146–47, 148, 150, 166; commenting system of, 176. *See also* NicoNico Chokaigi; NicoNico Cho Party
"Nobody Wins" (John), 108, 110
Numan, Gary, 92

objectivity, 16
"ODDS&ENDS" (Supercell), 150–51, 153, 154*fig*.
Off the Wall (Jackson), 126
Omura, Kenji, 99
OngaCrest (Core Research for Evolution Science and Technology), 145–46
On the Sensations of Tone as a Physiological Basis for the Theory of Music (von Helmholtz), 71–72
optimism. *See* technological optimism
Orbison, Roy, 157, 163
Orchestral Manoeuvres in the Dark (OMD), 91, 92
Organisation zur Verwirkichung gemeinsamer Musikkonzepte (Organization for the Realization of Common Music Concepts), 82
Ortega y Gasset, José, 86–87
Otis, Shuggie, 114, 115

"Packaged" (kz), 144
Paris World's Fair (1900), 10
Parker, Kevin, 13
Parkins, Elizabeth, 37
Pattie, David, 86, 92
Phillips, Simon, 116–17
phonographs, 10–11, 36, 55, 194, 195
Pianola, 31, 201n23; first advertisements for, 32

piano lessons, 28
piano-player, the, 22, 24, 26, 113; assessment of player piano technology, 32–33; difference of between piano-players and player pianos, 25; in early advertisements and articles, 30–34, 36, 38, 41, 42, 53; initial marketing of as automatic and self-playing, 24; as a labor-saving device, 26–30; movement away from its original identity, 47–48; push-up piano player cabinets, 24; and the term "player piano," 23–24, 24–25; as a threat, 34–40
Piano Player and Its Music, The (Newman), 41
piano rolls, 16, 20, 22, 22*fig.*, 26, 45–46, 69, 76, 79, 194; reproducing piano rolls, 52–53, 73–74
pianos: double-escapement piano action, 9; grand pianos, 7, 9, 58; as one of the most complex mechanized instruments in the musical world, 28. *See also* piano player, the; piano rolls; player pianos; reproducing pianos
Pinch, Trevor, 142
Player Piano (Vonnegut), 19
player pianos, 11, 19, 21–22, 26, 53–55, 192, 194, 195; difference of between piano-players and player pianos, 25; growing popularity of and the conflict that ensued, 35–36; improvement of player piano technology, 42–43; mastering of, 43–47; representations of in popular culture, 19–20; self-contained player pianos, 41; technology of, 3
pop music, 4, 8, 9, 11, 132, 166, 168
Porcaro, Jeff, 108, 109, 118–19, 119–20, 121
power, redistribution of, 190–91
Prince, 129–30; aesthetic goals of, 130

Recording Industry Association of Japan, 149

reproducing pianos, 4, 5, 11, 26, 53–55; advertisements for, 57–60, 58*fig.*, 61, 73; catalogs of reproducing piano rolls, 56; endings and beginnings of, 55–57; pianists that recommended the reproducing piano, 74–79; reproducing piano companies' focus on the "almost-presence," 60–61; reproducing piano rolls, 52–53, 73–74; selling and marketing of, 68–74. *See also* Welte-Mignon reproducing piano
Rhythmicon, 111–12
Riskin, Jessica, 5
Robinson, John, 126–27
robots/robotics, 11–12, 83, 84, 90–91, 181, 197–98. *See also* automata; Kraftwerk, robotic performance aesthetics of; "Robots, The" (Kraftwerk single release and video), reception history of
"Robots, The" (Kraftwerk single release and video), reception history of, 92–96, 95*fig.*
rock music, 11, 87, 90
Roland TR-808, 128
Roland TR-909, 128
Roland's Transistor Rhythm series, 114
Romanticism/romantics, 86–87
Rousseau, Jean-Jacques, 5–6, 54, 111
Ryan, William F., 12
"Rydeen" (YMO), 106
Ryo, 149–50

Sakamoto, Ryuichi, 96
salespeople, 41, 43
Sanden, Paul, 167
Sarah, Duchess of Marlborough, 40
Schedel, Margaret, 112
Schelling, Ernest, 74, 77–78, 79
Schneider, Florian, 85, 89, 197
Seaver, Nick, 71; on the rhetorics of fidelity, 72
"self-playing" instruments, 3–4, 30
"Senbonzakura" (Kurousa-P), 156
sequencers, 4, 95–96
Shibuya, Keiichi, 193

Side Man, 113, 114
Silicon Valley, 189
Sleeper Awakes, The (Wells), 10
Slonimsky, Nicolas, 112
Smith, Bertram, 46–47, 59–60, 192
Society for the Rehumanization of American Music, 6
Sohmer & Co., 42–43
Solid State Survivor (YMO), 96
Songrium, 145–46, 147*fig.*
Soul Train, 99–101, 100*fig.*, 103, 106, 207n43
Sousa, John Philip, 6, 11, 36, 40, 191, 194–95
Space Invaders, 103
Stilwell, Robynn J., 131
Stoddard, Charles Fuller, 65–66, 68, 71
Stravinsky, Igor, 21, 49
Strothmann, E., 36
Studies for Player Piano (Nancarrow), 112
Summer, Donna, 101
Sumrall, Henry, 98
Supercell, 149–55
synthesizers, 4, 88, 95–96, 106, 197–98; Minimoog synthesizer, 206n11; Moog synthesizers, 99. *See also* DX synthesizer series

Takahashi, Yukihiro, 96, 98, 99–100
Tame Impala, 13
Tangerine Dream, 83, 89
technological anxiety, 4, 10, 12, 13, 17, 36, 39, 40, 41, 82, 97, 116, 120, 123, 127, 189–91, 193, 195–96; anxiety of amateur music lovers, 109; anxiety concerning a German technological takeover, 85; hidden benefit of, 194
technological optimism, 3, 10, 12, 116, 126
technology, 7, 26, 176; "identity crisis" of, 18, 32; new music technologies, 194; relationship between technology and human identity, 10. *See also* automation/automation technology; music technologies; technological anxiety; technological optimism

Theremin, Leon, 112
"This Will Play Your Piano," 30
Thomas, Margaret, 37
Thousand Knives of Ryuichi Sakamoto (Sakamoto), 96
Tito, 185
Toffler, Alvin, 82
"tone test" concerts, 11
"Tong Poo" (YMO), 106
Tresch, John, 35
"Tried Dancing" (NicoNico Douga), 140
"Tried Singing" (NicoNico Douga), 140
Triumph Player Piano, 36–38
Trocco, Frank, 142
Tron (1982), 12
Tron: Legacy (2010), 13
Tubeway Army, 91–92
tune sheets, perforated, 29
Twitch, 138

Valerio, Angelico, 66
Vaucanson, Jacques de, 4, 34, 54, 133; defecating duck designed by, 4–5, 65; Rousseau's comments on, 5–6; "self-moving machine" designed by, 4
Vernon, Justin, 13
Victor Talking Machines, 55
violin players, 29
"virtual idols," 164
vocal coaches, 213n32
Vocaloid, 6, 13, 15–16, 17–18, 134–35, 157–58, 160, 190, 191, 193; absence of scholarship concerning, 162; and the barrel organ, 181; concerts of, 161, 165–66, 168–69, 176, 181–82, 185; criticism of, 158, 161; fans of, 140–41, 158, 160–62, 164, 165, 176, 185; first version of, 134; identity of, 142–43; popularity and influence of, 158; scholarly work concerning, 159; Vocaloid software, 154–55; Vocaloid songwriters, 156; Vocaloid videos, 160–61. *See also* Hatsune Miku; Vocaloid community
Vocaloid community, 136–38, 142, 146, 148, 149, 150, 154, 211n18

"VOC@LOID in Love" (OSTER project), 143
Voder (voice operation demonstrator), 133
Volavy, Miss, 59, 61
Von Helmholtz, Hermann, 71
von Kempelen, Wolfgang, 4, 133
Vonnegut, Kurt, 19
Vuitton, Louis, 193

Walden, Narada Michael, 126
Watanabe, Kazumi, 99
Wells, H. G., 10
Welte, Michael, 56
Welte Legacy of Piano Treasures, The, 64
Welte-Mignon reproducing piano, 2–3; advertisements for, 61, 63; early years of, 62–67; response to the performance of, 3; secrecy of, 63
Westworld (movie [1973]), 12
Westworld (TV series [2016–22]), 13, 19–20
"When Doves Cry" (Prince), 129
Wilcox & White, 30
Wonder, Stevie, 101
"World is Mine" (Ryo), 166
Writers Guild of America, 188
Wurlitzer, 113

X∞Multiplies (YMO), 103

Yamaha Corporation, 17, 134; synthesis technology of, 138
Yamaha DX7 keyboard, 166
Yellow Magic Orchestra (YMO), 11, 17, 80–81, 100*fig.*, 136, 205–6n3; album cover art of, 101–2, 102*fig.*; and *Circus*, 103–4; drum kit of, 98–99; early history of, 97–98; performance on *Soul Train*, 100–101; popularity of, 99, 107; reinterpretation of synthesizer performance by, 103–6; and technology and identity in eastern techno-pop, 96–102
YouTube, 13, 138, 139, 141–42

Zaborowski, Rafal, 161, 167
Zappa, Frank, 125

Founded in 1893,
UNIVERSITY OF CALIFORNIA PRESS
publishes bold, progressive books and journals
on topics in the arts, humanities, social sciences,
and natural sciences—with a focus on social
justice issues—that inspire thought and action
among readers worldwide.

The UC PRESS FOUNDATION
raises funds to uphold the press's vital role
as an independent, nonprofit publisher, and
receives philanthropic support from a wide
range of individuals and institutions—and from
committed readers like you. To learn more, visit
ucpress.edu/supportus.

www.ingramcontent.com/pod-product-compliance
Lightning Source LLC
Chambersburg PA
CBHW020809230426
43666CB00007B/927